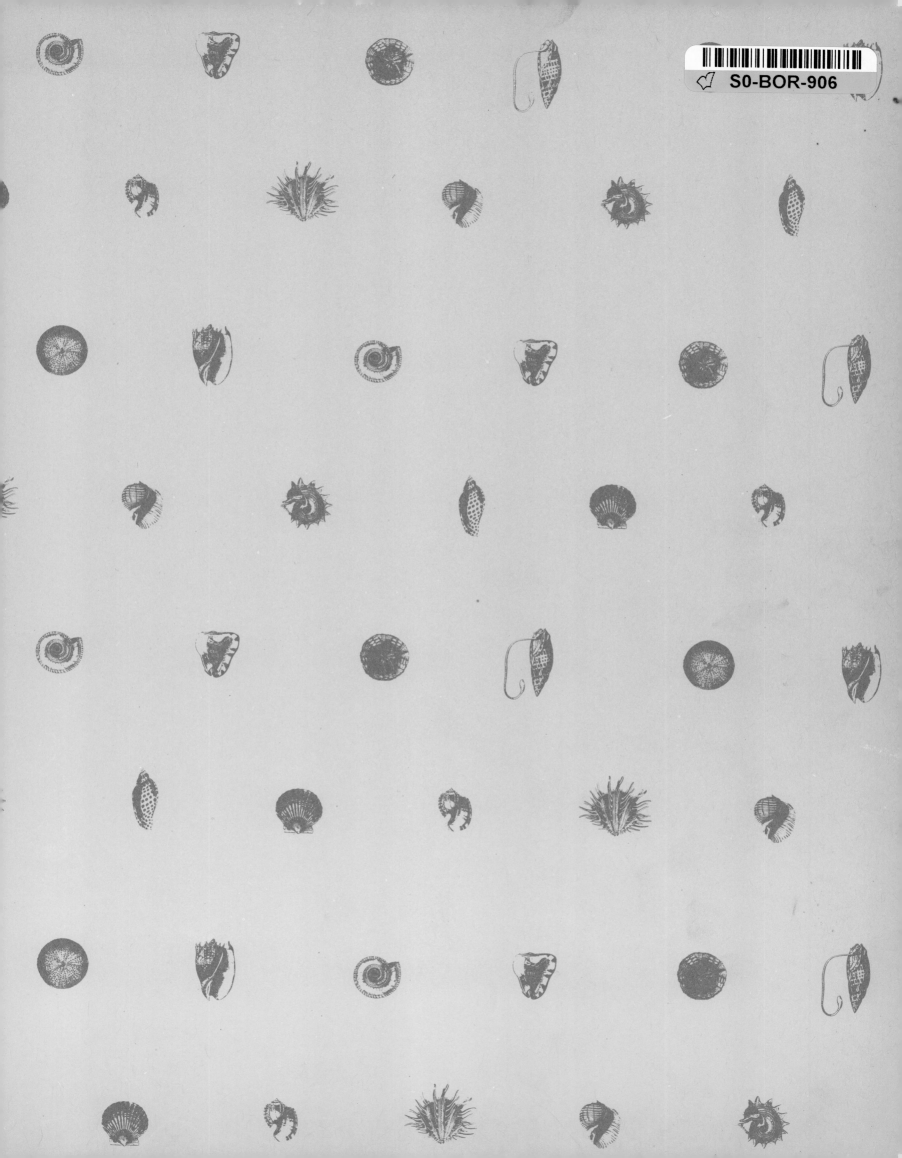

OCEAN LIFE

Discovering the World Beneath the Seas

Marty Snyderman
with
The International Oceanographic Foundation

PUBLICATIONS INTERNATIONAL, LTD.

Marty Snyderman is a world-renowned marine photographer and cinematographer. His work has been featured by the Columbia Broadcasting System, the British Broadcasting Corporation, the Audubon Society, the National Geographic Society, *Nova*, *Newsweek*, and the National Wildlife Federation. His writing and photography have appeared in a variety of publications including *Ocean Realm* and *Sea Frontiers*. A member of the Underwater Photographic Society, Snyderman has been working in the field for over 15 years. Typically, he participates in six to eight major marine expeditions per year. His excursions have taken him from the Sea of Cortez to the Red Sea and from the Galápagos Islands to Australia's Great Barrier Reef. He is affiliated with C&C Travel as their photographer and spokesperson.

Special acknowledgement is extended to Diane Rielinger and Lisa FitzGerald for their helpful editorial assistance. Both have masters degrees from the Rosenstiel School of Marine and Atmospheric Science and are on the editorial staff of the International Oceanographic Foundation's *Sea Frontiers*.

Cover photos:
Background: A school of yellow blue fish.
Front cover inset: Butterfly fish. *Back cover, top:* Fur seals. *Bottom left:* Cuttlefish.
Bottom right: Coral reef near Belize.

Copyright © 1991 Publications International, Ltd. All rights reserved. This book may not be reproduced or quoted in whole or in part by mimeograph or any other printed or electronic means, or for presentation on radio, television, videotape, or film without written permission from:

Louis Weber, C.E.O.
Publications International, Ltd.
7373 North Cicero Avenue
Lincolnwood, Illinois 60646

Permission is never granted for commercial purposes.

Manufactured in Yugoslavia.

8 7 6 5 4 3 2 1

ISBN 0-88176-964-9

Library of Congress Catalog Card Number: 91-61370

Photo credits:
Animals Animals: Doug Allan: 111, 277, 279, 280, 290; M.A. Chappell: 194; Margot Conte: 214; E.R. Degginger: 250; Steve Earley: 183; Ashod Francis: 294; John Gerlach: 201; Mickey Gibson: 36; Scott Johnson: 96; Richard Kolar: 207, 215; Zig Leszczynski: 61, 107; Bates Littlehales: 250; C.C. Lockwood: 203, 295; Tony Martin: 282; T.S. McCann: 280; Joe McDonald: 204; Stephen Mills: 132; Oxford Scientific Films: 314; Peter Parks: 299; Ralph A. Reinhold: 289; Carl Roessler: 162, 163, 170, 175, 182, 236; David Shale: 314; John Stern: 205, 217; Lewis Trusty: 249, 254, 256; James D. Watt: 165, 173, 208, 297; Bruce M. Wellman: 210; Anne Wertheim: 85; Fred Whitehead: 168; **FPG International:** 19, 34, 84, 125, 169, 235, 248, 276, 292; M. Alicia/Jill Wallin: 71; Laurance B. Aluppy: Back cover, Lee Foster: 196; Richard Gassman: 288; Giampiccolo: 53; John E. Gorman: Front cover, 12, 35, 57, 76, 77, 106, 108, 123, 140, 146, 148, 159, 229, 232, 239, 245, 246; Keith Gunnar: 51; H. Hall: 166; Richard Harrington: 194, 281, 293; Randall Hoover: 112; George Hunter: 284; M.P. Kahl: 289; Lee Kuhn: 221; Stephan Meyers: 48, 55; P. Millan: 62; Ernet Manewal: 274; Stan Osolinski: 105; K. Reinhard: 240; Carl Roessler: Front and back cover, 8, 22, 28, 31, 34, 40, 49, 70, 74, 87, 89, 99, 122, 125, 139, 140, 157, 158, 222, 223, 234, 235, 301, 307, 311; Leonard Lee Rue III: 188, 189, 190; Angabe A. Schmidecker: 142; Herbert Schwarte: Back cover, 35, 44, 225, 236, 237; Clyde H. Smith: 191; Karl & Jill Wallin: Contents, 4, 24, 73, 108, 114, 269; **International Stock Photography:** George Ancona: 63, 101, 102, 297; Dennis Fisher: 258; Tom & Michele Grimm: 17, 107, 129, 130, 145, 164, 184, 189, 196, 197, 206, 290; Steve Lucas: 79, 129, 143, 187, 223, 238; Donald L. Miller: 240; Jeff Rotman: 33, 135, 155, 224, 228; Robert Russell: 275, 285, 287; Robert J. Stottlemyer: 44, 100, 291; Bill Thomas: 286; Suzanne A. Viamis: 9, 211; **Marty Snyderman:** 6, 9, 10, 11, 13, 15, 16, 18, 20, 23, 25, 26, 27, 29, 30, 39, 41, 42, 43, 45, 47, 50, 52, 54, 56, 58, 59, 61, 64, 69, 71, 72, 77, 80, 85, 86, 88, 90, 91, 93, 95, 97, 99, 100, 103, 106, 109, 113, 116, 118, 121, 122, 124, 127, 128, 131, 133, 134, 136, 141, 144, 149, 150, 151, 154, 156, 157, 158, 160, 161, 167, 168, 171, 172, 174, 176, 177, 178, 179, 180, 184, 185, 186, 190, 192, 193, 195, 202, 212, 213, 219, 220, 226, 229, 230, 233, 249, 252, 253, 257, 259, 261, 263, 264, 266, 271, 283, 295, 298, 302, 304, 305, 308, 309, 310, 311; **Tom Stack & Associates:** 130, 137, 242, 315; Paulette Brunner: 19; Gerald & Buff Corsi: 8; David B. Fleetham: 38, 82, 95, 103, 110, 137, 141, 174, 218, 231, 300, 301, 313; Jeff Foott: 200, 215, 216, 283; Cindy Garoutte: 32, 48, 126, 243; Larry Lipsky: 33, 37, 142, 154, 247; Gary Milburn: 21, 99, 115, 167, 257, 268, 274; Mark Newman: 198, 199, 205, 273; Brian Parker: 12, 63, 92, 98, 106, 120, 138, 180, 241, 263, 270; Edward A. Robinson: 46, 104, 119, 147, 148, 181, 209, 210, 237, 246, 308, 312; Carl Roessler: 41; Jack Stein-Grove: 113; W.M. Stephens: 305; Jack Swenson: 218; Denise Tackett: 75, 187, 296; Larry Tackett: 43, 81, 243; Bill Tronca: 86; Stuart F. Westmorland: 14, 83, 110, 126, 244, 251, 255, 262, 265, 267, 268; Dave Woodward: 117; Anna E. Zuckerman: 272, 278, 292; **Norbert Wu:** 82, 273, 303, 306.

CONTENTS

Preface 4

Foreword 6

Introduction 8

To Live in the Sea 10

The Villains and Their Undersea Arsenals 42

Invertebrates 62

Vertebrates 124

Sharks and Their Kin 156

Marine Mammals 188

Life in Coral Reef Communities 222

Life in Temperate Seas 248

Life in Polar Regions 272

Life in the Open Ocean 294

Glossary 316

Index 318

A reef scene in the Philippines.

PREFACE

The International Oceanographic Foundation was established in 1953 to encourage oceanic research and cultivate new fans and friendships for the sea. The original Board of Trustees consisted of outstanding scientists and lay people who shared a deep devotion and respect for the ocean, and that tradition has continued to this day. With members in over 90 countries, the Foundation is one of the world's largest organizations devoted to spreading knowledge about the oceans. Focusing on life in the oceans, commercial fisheries, ocean sports, the seafloor, ocean currents, submarine and underwater habitat development, and other fascinating and burgeoning fields of marine science, the Foundation strives to expand and encourage humanity's understanding of nature's most prolific gift, the oceans.

Though independent for most of its history, the Foundation is now affiliated with the Rosenstiel School of Marine and Atmospheric Science at the University of Miami. Since 1943, the Rosenstiel School has been one of the premier marine education and research facilities in the world. The school is well known for its expertise and achievements in satellite oceanography, atmospheric chemistry, tropical studies, underwater acoustics, biology, and other areas of marine science.

For 37 years, the Foundation has produced the magazine *Sea Frontiers*, which devotes itself to fostering appreciation of the oceans and often publishes articles on topics covered by *Ocean Life*. Because of our shared interests and enthusiasms, the editors of *Sea Frontiers* happily lent an editorial hand to the book. The magazine staff includes graduate student researchers who check the factual accuracy of every story in *Sea Frontiers*. Two of those researchers, Diane Rielinger and Lisa FitzGerald, gifted marine scientists and graduates of the Rosenstiel School of Marine and Atmospheric Science, proffered their skills to help guarantee the accuracy of *Ocean Life*.

The author, Marty Snyderman, has contributed photographs to *Sea Frontiers* for many years, and our staff has always been impressed with the high caliber of his work. We are proud to be a part of *Ocean Life*, a project that so clearly achieves the Foundation's objectives and quality standards.

Bruce R. Rosendahl

Bruce Rosendahl
Vice President
International Oceanographic Foundation
and
Dean
Rosenstiel School of Marine and Atmospheric Science
University of Miami
4600 Rickenbacker Causeway
Miami, FL 33149

A sea turtle passes through a school of fish.

FOREWORD

The pull of the ocean is at once perfectly rational and bewilderingly emotional. Dedicated ocean lovers are drawn toward the sea as much by its secrets and mystery as by what can be learned and gained from studying it. We build diving machines, diving suits, and underwater habitats to try to overcome the basic inhospitality of the underwater world. The chief goal of all this inventiveness is to catch a better view of exactly what life is in the sea.

Ocean Life is a book about the cycles of life in the oceans, but it is loosely organized around food chains and predation. The book is defined by Marty Snyderman's unsparing love of the underwater environment. In the pages that follow, he offers a highly informed tour of the ocean's biology. Readers journey up the food chain from plant and plankton to invertebrates, vertebrates, and marine mammals. The trip is fresh and entertaining, and it includes some of the most fascinating, bizarre, and exciting stories from the marine world.

This book contains dazzling images from some of the world's finest underwater photographers. Drawing on his background as a marine photographer and writer, Snyderman offers essays on the richest ocean gifts. He has organized his subject in two ways: by class in the animal kingdom and by habitat. He has paid exceptional attention to how they kill, how they avoid being killed, how they breed, how they move, and how they depend on one another.

Clearly, Snyderman has great admiration for the creatures at the top of the food chain, the lively and sometimes dangerous sharks, rays, sea lions, whales, sea snakes, eels, and barracuda. But he reminds those who pick up this book that if they stop after studying only the marine animals with well-known reputations, they are missing a lot. What is at sea is rich, odd, and more amusing than is obvious. In truth, the real strength of this book is Snyderman's presentation of some rarely noticed creatures. He takes particular care with hole-dwelling organisms, the kind most photographers—and divers—don't stop for. He considers the small, inconspicuous, often overlooked ocean residents with the same respect and admiration he shows for great white sharks and whales.

Turn these pages and enjoy an underwater world that comes to life so clearly you might forget you're breathing air.

Bonnie Bilyeu Gordon

Bonnie Bilyeu Gordon
Editor
Sea Frontiers

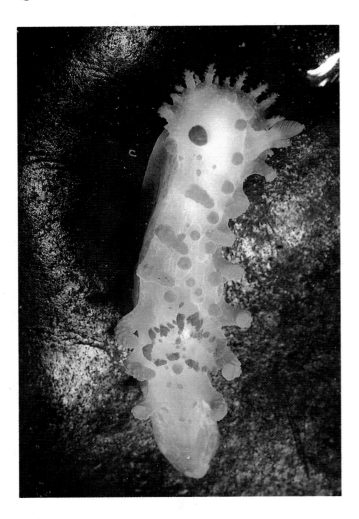

INTRODUCTION

Earth is truly the water planet. More than 70 percent of its surface is covered by water, and within those liquid boundaries live more than 200,000 different known species of plants and animals. These creatures range in size from microscopic one-celled plants that are the very foundation of much of the life on earth to spectacular 120-foot, 120-ton blue whales, the largest creatures alive today.

Truly, the oceans are not the empty voids they were once believed to be. Nor are they the dreaded domain of horrible sea monsters as they are so often portrayed in the sea lore of centuries past or modern Hollywood films. Instead, the ocean realm is a world of wondrous creatures whose amazing variety and marvelous adaptations cannot help but fascinate the newcomer and the most learned scientist alike. In many respects the allure of the ocean is that the more you learn, the more you realize how little you actually know.

Ocean Life is a journey inside this ocean realm. Its goal is to explore the marine wilderness and share the natural history of the exotic creatures that live in the sea. In essence, readers can glimpse the thrills and wonderment that only scuba divers can really understand.

Perhaps the most incredible of the ocean's treasures is the unimaginable variety of its plants and animals. Each of the seas' inhabitants fits into nature's scheme in its own particular way. There are similarities between many plants and animals, yet each species has adopted a survival strategy and form that is unique to itself. Despite their differences, all marine creatures are interrelated in some way. Each species depends upon a variety of others for food or protection and sometimes even for its own reproduction. Nature's craft and ingenuity are nowhere more evident than in the delicately spun web of life held within the oceans.

In *Ocean Life*, you will find spectacular images of many of the seas' most exciting animals, and you will learn about their various lifestyles. Intended for the lay person, this book is organized simply and clearly to provide an overview of the marine world, to offer a close look at individual creatures, and above all to show how the various species interrelate with each other.

The book first explores the unique challenges presented by the marine environment and the way its residents overcome these problems. From there, we enter the world of predators, from the hideously brutal sharks to the petite and lovely lionfish that kills just as efficiently.

The journey then moves through the scientific families of marine creatures, beginning with the primitive sponge and moving up the evolutionary ladder to vertebrates. Special chapters are set aside for two of the most interesting groups—the often misunderstood sharks and the lovable marine mammals such as seals, walruses, sea otters, and whales.

Finally we explore some specific marine habitats—coral reefs, temperate seas, polar regions, and the open sea. Each contains its own species of animals, specialists that are adapted to the particular conditions of their environment. Each houses its own brand of mystery, drama, and beauty. And each is a place like nowhere else on earth.

Opposite page, top: *This sea clown nudibranch could easily be mistaken for some type of worm. Actually, it is a mollusk, closely related to the snail and the octopus. Nearly all nudibranch species are fantastically colored, and they can be found in all major marine habitats.* Bottom: *A bright yellow butterfly fish darts about a coral reef. It is one of thousands of species of fish that thrive in tropical reef communities.* This page, top: *A ray glides through the water with unmatched grace and elegance. Many species of rays make their homes in reefs or other confined habitats, while some live their entire lives in the expansive open ocean.* Bottom: *A California sea lion rests on a rocky shoreline. Seals and sea lions are highly social mammals that have adapted their bodies and their behavior to the marine environment.*

TO LIVE IN THE SEA

The life history of all creatures is a battle for the survival of their species. Marine animals are no exception to this rule. The natural life cycle of every fish, dolphin, sponge, and sea star focuses on surviving long enough to produce offspring that will perpetuate the line. They must evade heavily armed predators and capture well-defended prey long enough to reach sexual maturity, reproduce, and ensure that some offspring survive.

How the various species survive in the sea long enough to perpetuate their own kind is a fascinating study and a great marvel of nature. Every species takes a slightly different path or uses different adaptations and strategies to survive and produce young. This is what characterizes the species and determines how and where each will live or die.

Nature shows its generosity in the numberless talents and tricks creatures use to survive the harsh marine environment. The barracuda uses its incredible speed, while the stonefish and octopus are masters of camouflage. Sharks can sense electrical fields generated by other living organisms, a skill that enables them to

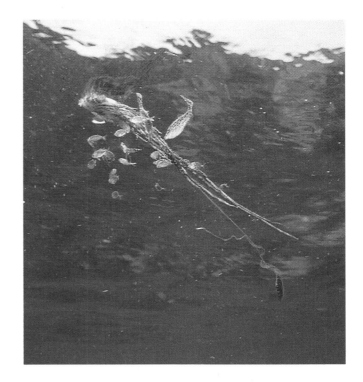

Opposite page: *A school of Pacific barracuda prowl the waters off the Galápagos Islands. The barracuda can accelerate from a dead stop to full speed almost instantly, making it an extremely capable hunter.* Above, left: *The parrotfish's jaws each contain a number of individual teeth fused together to form the distinctive beak.* Above: *A Portuguese man-of-war ensnares a young filefish. Its tentacles contain tiny stinging cells that release deadly toxins into any creature that wanders into them. A close relative of the true jellyfish, the man-of-war can grow tentacles to well over 100 feet long.*

TO LIVE IN THE SEA

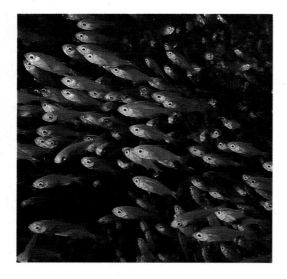

This page, top: *The brittle star's arms consist of a series of many separate segments, giving the arms their sinuous, snakelike movement and making the creature very maneuverable.* Above: *Glass fish swarm together near the Maldive islands in the Red Sea. Dense schools such as this one increase the likelihood of each individual's survival.* Opposite page: *Perhaps nature's ultimate predator, the great white shark is designed to hunt, from the well-developed sensory organs in its snout to its powerful, thrusting tail.*

detect well-hidden prey. Corals, sea anemones, and jellyfish use deadly stinging cells to hunt and to protect themselves. Parrotfish teeth fuse into a birdlike beak for biting and chewing coral and rock to get the nutritious algae found within. Goatfish have chemical receptors on the ends of their whiskerlike barbels that help them locate worms and crustaceans in the sand.

Some species seem to depend heavily on one particular skill, but most animals employ a combination of adaptations in their quest to capture food and survive an ever-changing environment. Sharks, for example, are famous for the rows and rows of razor sharp teeth most species have. But great white sharks, blue sharks, makos, hammerheads, and the other great hunters need more than sharp teeth to secure their food. Nature also gives them protruding jaws that extend forward when the mouth opens, so the teeth are in the best position to grasp large prey. Sharks also have well-developed sensory systems to help them find their prey, and their streamlined shape and great power give them the speed to pursue their meals.

Predators such as barracuda, jacks, dolphins, and killer whales are among the fastest swimmers, but not every animal can be fast. Those that are not need other adaptations. Brittle stars, for example, are not particularly fast, but they are highly maneuverable. They evade rather than outrun their enemies.

Grunts, snappers, mackerels, and other fish gather in large, organized groups called schools. Schooling makes them less vul-

TO LIVE IN THE SEA

TO LIVE IN THE SEA

nerable to predators and increases their chance of finding food and mates.

Some fish don't rely on schooling for protection. By swallowing water, spiny puffer fish inflate themselves, erecting spines all over their bodies to deter their foes. Lionfish and stonefish defend themselves with toxic spines in their dorsal fins. Other animals, too, rely on their outer coverings. Long, sharp spines encase some sea urchins' soft, edible bodies, while lobsters, crabs, and turtles find safety in their hard shells.

Angelfish and other species that are not so well protected are adept at slipping into small crevices in a reef. But no one in nature has a free ride. Predators like moray eels are well designed for hunting in the caves and crevices where many species hide. Another crevice resident, the octopus, can counter the eel's threat by releasing an inky fluid that clouds the water and screens its escape.

Some creatures use camouflage to avoid predators and improve their own hunting ability. Flatfish like turbot, halibut, and sole alter their coloration and patterning to blend in to almost any background, as do octopi. Most stingrays and other residents of sand communities that cannot change their color instead bury themselves to avoid detection. The stingrays' eyes rest high atop their heads so they can survey their surroundings as they hide. Stingrays also have spiracles, or openings behind the eyes that allow oxygen-rich water to reach their gills beneath the sand.

The comet fish, or roundhead, uses a remarkably clever mimicry to protect itself. Its pattern of small white dots on a dark background is very similar to that of the white-speckled moray eel. It also has a large dark spot in the rear and a distinctively tapered tail. When threatened, the roundhead hides the front half of its body in a hole or crevice. The protruding tail looks almost exactly like the head of an eel peering from its den, a sight that will discourage many predators.

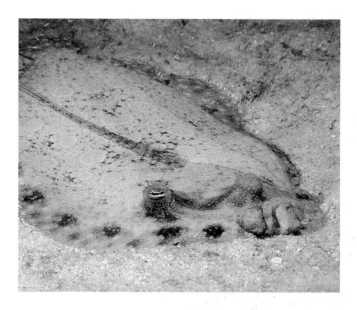

Opposite page: *Threatened by a diver, this giant Pacific octopus releases an ink screen to cover its escape. The ink may also act chemically to dull some predators' sense of smell.* Above: *The peacock flounder uses its flattened body shape and ability to change colors to become almost invisible to both predators and prey.*

TO LIVE IN THE SEA

Surprisingly, the bright colors and patterns of many tropical reef fish make it difficult for their predators to see their overall shape. Foureye butterfly fish use two large false eyespots near the tail to confuse predators about which end of the fish is which. The predators can't anticipate which direction the fish will swim as it flees an attacker.

Nudibranchs are mollusks that have no shells. Many have soft, brightly colored bodies that are easy to spot. Over the millennia, some nudibranchs have developed a body chemistry that simply makes them taste bad to other animals and so ensures their survival. Some nudibranchs have also learned to steal the stinging cells from hydroids and corals and use them in their own defense.

Above: *Delicate and beautiful, these nudibranchs have no way to hide or fight against predators. Their offensive taste is their only defense.* Opposite page: *This northern fur seal will spend months raising her young on the Pribilof Islands in the Bering Sea.*

In addition to their great speed and high intelligence, dolphins use echolocation to hunt and to avoid danger. Echolocation is a natural sonar that enables dolphins to analyze their surroundings. They emit high-pitched sounds and listen for changes in the sound waves caused by any bodies in the environment. Some studies show that dolphins may also use extremely high frequency sounds to stun potential prey.

Reproductive strategies are also crucial to a species survival. Many animals have remarkable traits that give their species a reproductive edge. For example, some fish such as wrasses can change sex in the middle of their lives according to the needs of local populations. Other animals such as many snails and nudibranchs are hermaphroditic, meaning they have both male and female reproductive organs. Jellyfish reproduce sexually in one generation, and asexually by budding, or splitting apart, the next.

Some invertebrate animals like sponges reproduce sexually even though the individuals have no physical contact. They release enormous quantities of sperm and eggs into the water. Through blind chance, a few eggs are fertilized and produce adult sponges. Octopi produce far fewer eggs, but the females spend a great deal of energy guarding and cleaning them. Some male fish like sergeant majors and garibaldis constantly guard their nests. Many jacks are broadcast spawners that mate in large groups. The males and females release their spawn into the water; the strategy is similar to that of sponges, but it's more precise.

TO LIVE IN THE SEA

Whales, dolphins, seals, sea lions, and other marine mammals tend their young for extended periods of time. As with all mammals, the females nurse their young. Yet sharks and many fish readily devour their own young. The mammals' strategy is to protect the precious few offspring they produce, while some fish and most invertebrates simply produce many young and hope a few will survive.

Reproduction also explains why many fish are so colorful. Scientists believe that colors and patterns can be a form of sexual communication. In some species males have different color patterns than females, and juveniles are different from mature adults.

No animal lives forever, and life spans are often cut short in the brutal ocean environment. Still, successful sea creatures have adapted so that members of each species can survive, reproduce, and avoid extinction. The behaviors that ensure each species' survival are cast in the genetic codes of the individuals, who pass the information from generation to generation.

OCEANIC FOOD CHAINS

All plants and animals in the world affect each other in some way. Although there are a great many kinds of relationships among animals, perhaps the most obvious and fundamental one exists between predator and prey. Scientists recognize the importance of this type of relationship. They use the term trophic to describe the feeding relationships among various organisms. Lay people often use the term food chain.

The marine environment, perhaps more than any other, clearly shows the complexity and scope of the earth's food chains. In the Antarctic seas, for example, a simple but definite food chain consists of plankton and baleen whales. Zooplankton (animal plankton) prey upon phytoplankton (plant plankton). In turn, various zooplankton are food for the filter-feeding baleen whales.

However, in nature things are never that simple. Such a straight line food chain is very rare. Other organisms get involved, and separate food chains become interwoven. In the Antarctic food chain, several other species come into play. Squid and some fish also feed on the plankton. Many larger fish prey on the filter-feeding fish and squid. Killer whales not only hunt these larger predatory fish but also attack the baleen whales. To further complicate things, the system extends out of the water. Penguins and seals, both land-based creatures, eat the squid and fish, and they serve as primary foods for polar bears and the killer whales.

Even this convoluted food chain is a relatively simple one. Through a variety of complex, multilayered relationships, eating habits link virtually all the creatures of the sea. There are literally thousands of interconnected food chains. Scientists refer to these interconnected chains as food webs.

In all tropical, temperate, and polar seas, plants form the foundation for most major food webs. These plants take raw material—sunlight and chemical compounds—and create organic matter and energy through photosynthesis. Plants are producers, or first-level members of the food chain.

The capture and use of energy from the sun by plants only begins the story. From plants, the energy flows toward consumers, first to plant-eating animals called herbivores. The kelp isopod, an animal that grazes upon kelp, is a herbivore. From the herbivores, the energy moves toward the carnivores. A rockfish, an animal that commonly feeds on kelp isopods, is a carnivore. From there, the energy transfers to still other carnivores, such as a salmon that preys on rockfish. Each step along this path of energy flow is called a trophic level.

Consumers cannot produce their own food from an inorganic substance such as sunlight. In one way or another, almost all animals depend on the plants for their existence. This dependence is equally true of tiny zooplankton and of great white sharks. Even the apex predators, the animals that sit at the top of a food chain, derive their energy from plants; the energy simply passes through many hands before it gets to them.

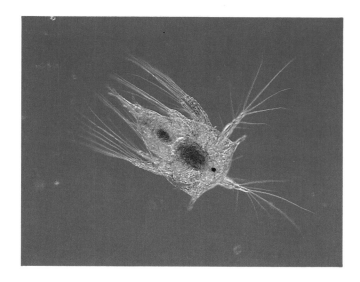

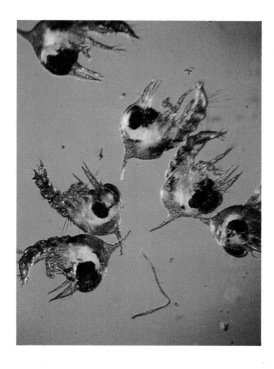

This page: *This barnacle larva (top) and these microscopic crustaceans (above) are part of the plankton that support marine food chains.* Opposite page, top: *The kelp snail is a marine herbivore. The barnacle attached to its shell is also feeding, drawing plankton from the water currents.* Bottom: *A sea star consumes a tiny red crab.*

TO LIVE IN THE SEA

Within all food chains, energy almost always passes in one direction; it moves up the chain from producers toward the top-level predators. However, the energy transfers very inefficiently. A great deal of waste occurs at every level. Much of the solar energy collected by a plant, for example, does not get stored as sugar molecules produced by photosynthesis. Some burns off during normal metabolic processes and some escapes as waste products.

Similar waste occurs in the herbivores and carnivores. Bodily functions from respiration to reproduction require energy. Swimming has an energy cost, and capturing food requires energy. And no animal can convert all the potential energy in its food into a usable form. These losses mean that each trophic level contains less usable energy than the level directly below it.

Studies show that as energy moves from one rung in a food chain to the next, only between six and 20 percent of it survives. Efficiency varies from consumer to consumer, but scientists use 10 percent as an average. A convenient but slightly inaccurate way to put this is that it takes 10 pounds of plankton to support one pound of squid and 10 pounds of squid to support one pound of penguin. An apex predator like the killer whale might be on the sixth or seventh level of a food chain. A small one-ton killer whale can require the energy of as much as a million tons of plants.

The complexities of food chains and energy transfer reveal much about life in the sea. Phytoplankton and zooplankton become vitally important to the marine kingdom. These creatures are not insignificant plants and invertebrates; instead they are the very foundation for much of the ocean's life. All living creatures, from the tiny phytoplankton to the top-end predators, from the tropical corals to the polar marine mammals, are dependent on their neighbors. The interwoven systems are complex, and though they show some resilience, perhaps the greatest lesson is that they are fragile. Every species plays a vital role in the web of life, and if any one species disappears, many are affected.

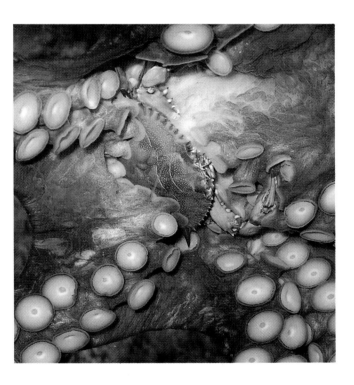

Opposite page: *A black grouper pursues a huge school of silversides in a reef near the Grand Caymans.* Above: *Using its powerful suckers to hold the prey and its beak to tear at the flesh, this octopus consumes a helpless crab.*

RELATIONSHIPS BETWEEN MARINE ANIMALS

The ocean demands close interaction among its inhabitants. The most obvious relationship between two different animals is that of predator and prey. But there are other relationships that do much to shape the ocean's daily life cycle.

For example, cleaning relationships are common in many different species. Certain types of shrimps and small fish like gobies and wrasses actually clean other creatures. The cleaners rid the other fish of parasites, dead tissue, loose scales, mucus, and bacteria while getting themselves a meal. This is often a mutually beneficial interaction, or mutualism, where both parties benefit and neither suffers.

In Caribbean reef communities, it is common to see gobies and several species of shrimps hard at work cleaning Nassau groupers, tiger groupers, coneys, rock hinds, and any number of other fish in a scene that looks like an underwater carwash. The cleaners posture and swim in distinctive patterns to advertise their established cleaning stations. When the cleaning activity is especially heavy, animals who want to be cleaned literally hover in line awaiting their turn. At times cleaners will actually swim into the mouths of the fish they are cleaning, even with dangerous predators like moray eels.

Usually the fish that want to be cleaned posture to let the cleaners know they want their services. This behavior is also a sign that the cleaners face no danger. Groupers often hover mouth agape and gills flared to get the cleaners' attention. Creole fish hang at a slight angle with their mouths open, heads down, and tails up. Triggerfish usually stand on their heads in a request for cleaning, while some surgeonfish stand on their tails. When they have had enough, they again perform a distinctive series of moves to let the cleaners know they should depart.

Most cleaners do not rely solely on their cleaning activity to gather food. And some species such as blueheaded wrasses are active cleaners only as juveniles. However, cleaners such as the cleaner wrasse found throughout Indo-Pacific reefs apparently use their cleaning skills as their sole means of nutrition.

Perhaps some of the most remarkable interactions among marine animals are cleaning relationships. This page: *A cleaner wrasse prepares to enter the jaws of a white-spotted moray eel. The fish will emerge not only unharmed, but with a full stomach as well.* Opposite page: *A group of Caribbean creole fish posture. Their stance shows that they would like to be cleaned.*

TO LIVE IN THE SEA

TO LIVE IN THE SEA

While mutually beneficial relationships are a good thing for all parties concerned, in nature there are always some species that will take advantage of a situation. The aptly named false cleaner and mimic blennies provide a classic example. These blennies look quite similar to cleaner wrasses. In addition, the blennies mimic the swimming antics of wrasses so well that many fish get fooled into believing the blennies will clean them. When they get their opportunity, the blennies are quick to bite a chunk out of the fins and skin of their surprised victims.

An important symbiotic relationship exists between host corals and one-celled algae that live within the coral tissues. The algae, called zooxanthellae, provide the corals with energy-rich compounds made during photosynthesis. The algae receive nutrients produced by the coral's metabolic process. This seemingly limited relationship plays a vital role in the innerworkings of every tropical reef community. Zooxanthellae serve as the very foundation of the reef's food chain, and the coral skeletons provide habitats for most reef dwellers.

Giant clams also enjoy a mutually beneficial relationship with zooxanthellae. The clams live in shallow, protected waters in conditions that are ideal for the algae. The algae live in the tissue of the clams, and their presence shows itself as the brilliant colors in the clams' mantles. The clams receive organic materials from the algae.

Below: *This masking crab is one of many crustaceans that hide from predators by adorning themselves with debris or anemones, sponges, and other animals.* Opposite page: *The mantle of this giant clam gets its brilliant hue from the algae that live within.*

TO LIVE IN THE SEA

TO LIVE IN THE SEA

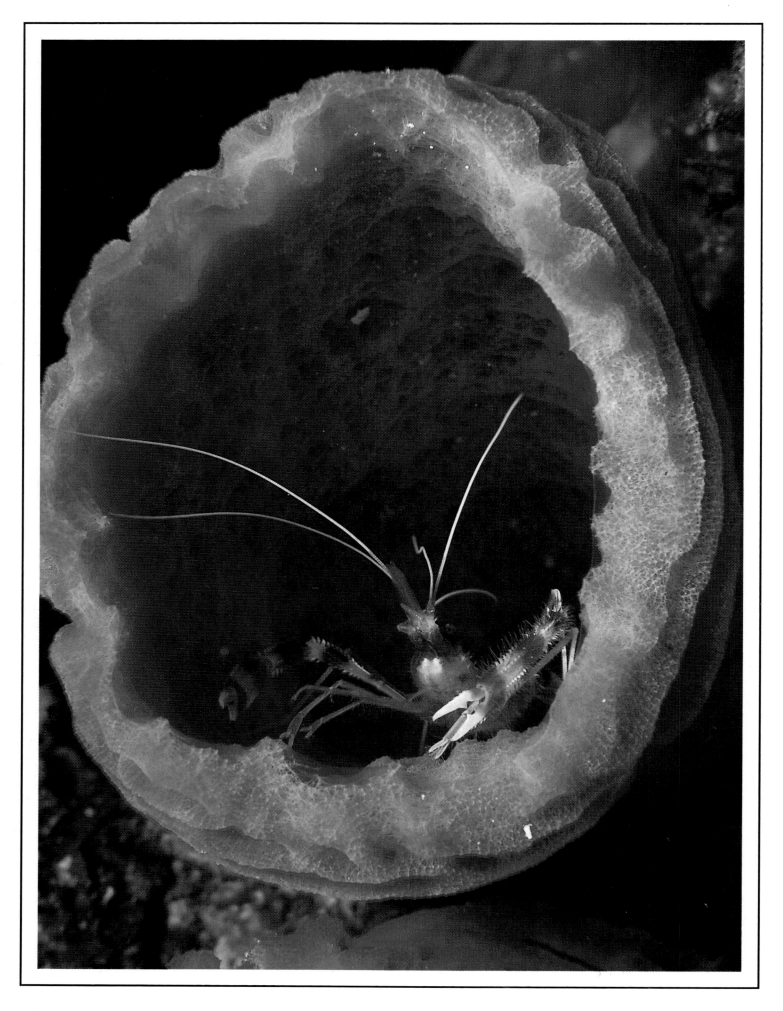

Several species of hermit crabs cover their shells with living sea anemones. The hermit crab entices the anemones to release their grip on the bottom so the crab can put them on its shell. The anemones provide the crab with a protective disguise, and the crab's mobility gives the anemones a better chance to gather food.

Some burrowing shrimps and certain gobies enjoy another mutually beneficial relationship. These animals live together in the same burrow, one that is dug by the shrimp. The burrow provides a convenient shelter for the goby, and the goby offers the shrimp security services in return. Standing guard at the top of the burrow, its tail always touching the shrimp's antennae, it watches for predators. When danger approaches, the goby retreats into the burrow, signaling the shrimp that it should hide as well.

Some mutually beneficial relationships are essential for one partner but only a matter of convenience for the other. Highly active anemonefish live unharmed among the toxic stinging tentacles of sea anemones, and both species benefit in several ways. The fish, for example, evade predators by hiding in the anemones' tentacles, and the anemone feeds on scraps from the fish's meals. Under normal conditions, however, anemonefish are always found near their anemones, but the anemones can survive quite well with or without resident anemonefish.

Scientists generally refer to interaction between two organisms as symbiosis, or symbiotic relationships. Mutualism is only one form of symbiosis. In commensal relationships one partner benefits while the other partner neither benefits nor suffers. In parasitism, one animal, the parasite, benefits while the host animal is harmed. Although true parasites do not kill their hosts, they do make life more difficult.

The relationship between some larval fish and species of long-spined sea urchins provides an excellent example of a commensal relationship. The fish finds safety from potential predators by hiding among the urchin's spines, while the urchin remains unaffected. Many species of shrimps, brittle stars, and crabs gain protection and food by hiding in sponges. The residents do not eat the sponge, but they are quick to grab any organisms that float past. The sponge remains virtually unaffected, with the possible exception that it loses some nutrients to its guests.

Opposite page: *A cleaner shrimp peers from the sanctuary of a sponge.* Above: *A goby and a blind shrimp share a burrow. Gobies will also take up residence with aptly named innkeeper worms, which also dig burrows.*

TO LIVE IN THE SEA

Parasitic relationships are common as well. Copepods and isopods are crustaceans closely related to crabs; some are parasitic. Sharks and many fish spend much of their lives with attached copepods on their fins and in their mouths. Isopods attach themselves to many species including garibaldis, angelfish, squirrelfish, soldierfish, and many more.

As with the gobies and groupers and the copepods and sharks, marine animals weave a complex fabric of life. No organism exists alone. Each plays its own crucial part in maintaining the balance of life in the sea.

Opposite page: *A tropical soldierfish unwillingly feeds a parasitic copepod.* Below: *The California spiny lobster may look ungainly, but it is actually very agile and quick.*

SHAPE AND LOCOMOTION IN MARINE ANIMALS

Snorkelers and divers can easily understand how sea lions, sharks, dolphins, and other large animals can outswim their human counterparts. It is far more difficult to see how a short, stubby boxfish or an ungainly lobster can handily outmaneuver skilled human swimmers. Marine animals have had countless thousands of years to live in their fluid environment. In that time, their bodies have adapted some ingenious techniques for moving about their unique world.

As land animals, we are most familiar with movement through a gaseous environment where we have little buoyancy, pushing air molecules out of the way and struggling against gravity. While we are well suited for locomotion on land, water presents a different set of problems.

TO LIVE IN THE SEA

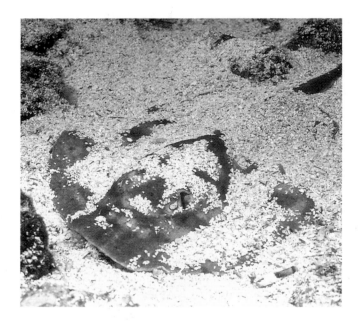

This page: *The stingray* (top) *and blue shark* (above) *both lack swim bladders to make them buoyant. Their body shapes and lifestyles have developed along different paths to counter this disadvantage.* Opposite page: *This angelfish hovers effortlessly over a reef in the Solomon Islands.*

The salt water of the ocean is about 800 times more dense than air is at sea level. This great density makes moving in water harder than moving in air. The biggest obstacle in a liquid environment is the drag created by the heavy water. Drag, or friction, on a marine animal is affected by the animal's shape, its speed, the structure of its surface, and the way water flows over that surface. As a general rule, the longer the animal's body and the more streamlined its shape, the smoother the water flow and the faster the animal can move. However, the faster an animal tries to swim, the greater the drag.

Still, sea creatures do enjoy some advantages by living in the water. Since water is denser than air, organisms are more buoyant in the liquid environment. An animal will float if its density is less than the density of the water.

Most bony fish such as damselfish, parrotfish, triggerfish, and butterfly fish also have an internal organ called a swim bladder that lets them float. Because of bones and other heavy body parts, an individual fish is more dense than the water that surrounds it, which means that the fish should sink instead of float. If the fish inflates its swim bladder with gas, however, its density decreases to match the surrounding water, and it can float effortlessly.

Sharks, rays, and skates lack an internal swim bladder. As a result they must expend considerably more energy to avoid sinking, or they must adapt to life on the seafloor. Angelsharks, guitarfish, horn

TO LIVE IN THE SEA

TO LIVE IN THE SEA

sharks, and stingrays have learned to use their lack of buoyancy as an advantage. These creatures use their weight to help keep themselves on the bottom even in heavy surge and strong currents.

Blue sharks, mako sharks, hammerheads, and manta rays have no swim bladder and must spend their entire lives swimming to keep themselves from sinking. These species have skeletons of cartilage, which is lighter than bone and makes constant swimming a little easier. Some species of bony fish such as tunas also lack a swim bladder.

Obviously fish occur in a wide variety of shapes. If you take a close look at the shape of any given fish, you can discover revealing insights into the fish's natural history and lifestyle. Their shape impacts where they can live and the way they pursue their food and escape predation.

Fish occur in four basic shapes. Fusiform species are more or less torpedo shaped. Most have a slightly rounded head and a body that tapers behind. Examples include barracuda, great white sharks, groupers, tuna, and lizard fish, all extremely fast swimmers. The bills of the marlin and sailfish create a modified fusiform shape. These species are also among the fastest swimmers. Some tuna have been clocked at 46 miles per hour. They cannot maintain such speed for more than a few seconds, but even an instant's speed can make the difference between escaping a predator's jaws and being eaten.

Typical of fusiform fish species, the barracuda (opposite page), *the open ocean jack* (this page, top), *and the sand tilefish* (above) *are all extremely fast swimmers.*

TO LIVE IN THE SEA

This page: *The batfish* (top) *and butterfly fish* (above) *are laterally compressed. The tall thin shape allows them access to areas of the reef that stouter creatures can't enter.* Opposite page: *The yellow stingray benefits from its dorsoventrally compressed body both when moving and when stationary. The shape lets it glide gracefully through the water and lie flat and unseen on the seafloor.*

Angelfish, butterfly fish, and spadefish have bodies that are laterally compressed; they are much taller than they are wide. They are not the fastest swimmers, but their body design affords them access to many small spaces. There they find both food and protection from their enemies.

Other fish have dorsoventrally compressed bodies, or ones that are much wider than they are tall. They spend the vast majority of their lives on or near the seafloor or gliding in the open sea. Stingrays, bottom-dwelling sharks such as angelsharks, and flatfish such as halibut, turbot, sole, and sanddabs are excellent examples. These species have their eyes located well up on their heads and can bury themselves in the sand with only their eyes showing. Hidden this way, they are better able to avoid potential predators and to surprise their prey. Manta rays, mobula rays, and spotted eagle rays spend their lives cruising in the open sea. Their flattened bodies make them superb gliders and long distance swimmers.

Members of the eel family have snakelike, or attenuated, bodies. These species maneuver easily in the restricted confines of rocky areas and coral reefs, making them formidable predators.

Most fish, no matter what their shape, propel themselves mostly with their trunk muscles and the motion of their tail. The head remains relatively still while the tail end pushes the fish forward. The most important function of the other fins is to help control turns, stopping, and hovering.

TO LIVE IN THE SEA

The structure of the tail is critical in many marine creatures. Many open ocean species such as blue sharks and hammerheads have long powerful tails that supply plenty of thrust and lift. The broad, flat tail design and heavy musculature of seabasses and groupers helps them accelerate rapidly, an important asset for predators in reef communities.

Cetaceans, the whales and dolphins, have tails designed to help them reach the surface and breathe. Like other mammals, whales and dolphins breathe air; they have no gills. Unlike all fish, the tails of cetaceans lie horizontally rather than vertically.

Some attenuated species: The moray eel (opposite page), *a group of needlefish cruising the surface waters* (above), *and a trumpetfish* (below).

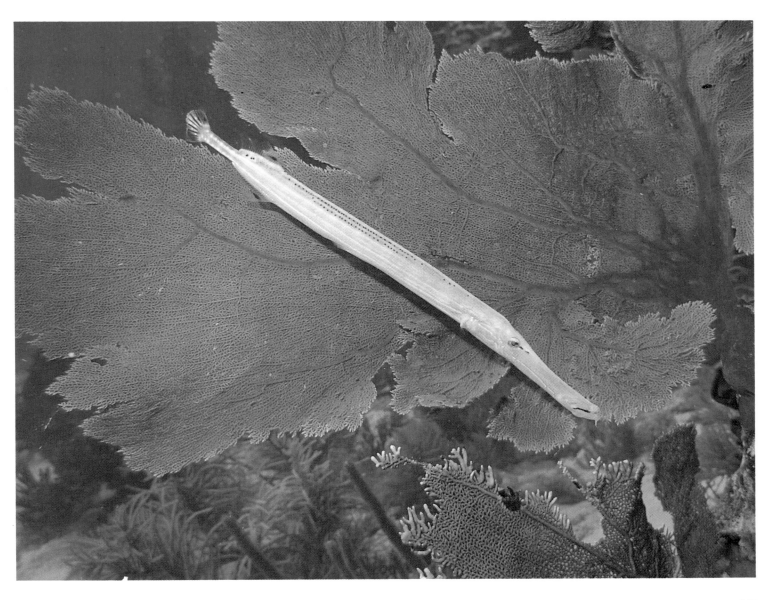

TO LIVE IN THE SEA

This design enables them to more easily thrust themselves toward the surface where they can capture a lungful of life-sustaining air. In addition, their blowholes, the openings through which these animals breathe, sit at the top of their heads. As a result, whales and dolphins do not have to hold their heads high out of the water. They only have to break the surface to expose their blowholes to air.

Below: *A bottlenose dolphin breaks the surface to grab a lungful of air.* Opposite page: *A right whale shows its broad horizontal tail.*

For years scientists and engineers have been intrigued by the high speeds attained by dolphins. Many species can swim over 20 miles per hour. Experiments show that a rigid model of a dolphin moving through the water at about 20 miles an hour generates more drag than a dolphin could overcome with muscle power alone. While dolphins do not have the strength needed to reach their top speeds, they do have the ability to alter their skin's shape as they swim to lessen resistance. They can do this so quickly that they greatly reduce the turbulence of water flow around them even as they swim at high speeds. Obviously greater speed is a tremendous advantage when pursuing prey or outdistancing predators.

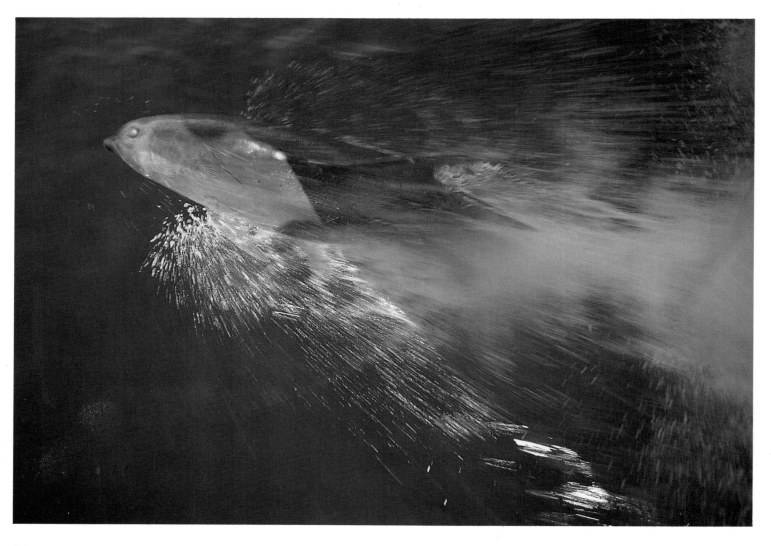

TO LIVE IN THE SEA

TO LIVE IN THE SEA

TO LIVE IN THE SEA

Perhaps the most efficient marine locomotion system is the jet propulsion used by octopi and squid. Octopi and squid propel themselves with a powerful stream of water. First, they draw the water into a special part of their body called the mantle cavity. Then they quickly force the water out a directable tube called a siphon. The animal moves in the opposite direction to which the siphon is aimed. At times octopi also form their bodies into a wing shape to create lift and become more efficient swimmers.

Even this quick study of shape and locomotion shows much about life in the marine environment. It's easy to see why torpedo-shaped fish like barracuda are such excellent predators, why laterally compressed butterfly fish are likely to remain close to the recesses of a reef, and why whales and other marine mammals have horizontal tails when no other sea creatures do. Perhaps more important, this knowledge also creates new respect for the ingenuity of nature.

Opposite page: *A moray eel darts from a crevice in the Red Sea to grab a meal.* This page, top: *An octopus glides over the seafloor on a jet of water.* Above: *This white-tipped reef shark relies on its streamlined shape and large, powerful tail to chase down prey in the Coral Sea.*

THE VILLAINS AND THEIR UNDERSEA ARSENALS

The sea is a place of startling beauty, delicate intricacy, and dazzling wonder. It is also a place of great savagery and danger. For creatures to survive in the ocean, they must win out over their competitors, their hunters, and their prey. Often the most direct course to survival means being more muscular, more toothy, more toxic, or just plain more deadly. Many ocean creatures have taken this evolutionary path. From sharks, barracuda, and venomous sea snakes to fire worms, corals, and sharp-spined urchins, the marine world houses its share of well-armed creatures.

People do not belong to the marine world. No ocean creature has evolved to be a direct threat to humans, and certainly none hunt humans as a part of their natural diet. In fact, nearly all marine creatures shy away from people. Still, many ocean inhabitants could kill or seriously wound a human being with ease. A well-educated diver knows that caution and distance are the best guarantees of safety.

Opposite page: *The jagged, razor-sharp teeth of a passing great white shark indicate the animal's carnivorous habits. Not all marine organisms, however, display their deadly potential so obviously.* Above, left: *Stinging cells, fatal to prey and often annoying to humans, lurk within the gentle beauty of a Pacific soft coral.* Above: *A well-camouflaged scorpionfish conceals its location and the array of poisonous spines in its dorsal fin.*

THE VILLAINS AND THEIR UNDERSEA ARSENALS

This page, top: *A moray eel will greedily snap up small reef animals, but it poses no real threat to scuba divers.* Above: *A killer whale glides through the water in search of prey.* Opposite page: *Although great white sharks have attacked humans, we are not part of their normal diet.*

THE PUBLICIZED PREDATORS

Mouth agape and exposing rows of razor-sharp teeth, the great sea creature silently approaches its victim. Within seconds, the marauder has overwhelmed its surprised prey. The water turns red as the duo disappear into the murk. Despite its frantic efforts, the victim never really had a chance against the skills of the superpredator.

And who was this killer? If this were a novel or movie, it could have been a shark, a barracuda, a killer whale, or one of the other sea creatures so often appearing this way. Certainly all these animals are able predators, and each must fend for itself in the world of nature. But much of their vicious reputation started in the myths of sea lore and became even more distorted as a part of popular entertainment.

When the topic is sea monsters, the first animal that comes to mind is the great white shark. Active predators, as adults they often hunt seals, sea lions, dolphins, and other marine mammals. They are also responsible for many verified attacks on people. Caught in Cuban waters, the largest documented great white shark was 21 feet long and weighed just over 7,000 pounds. The teeth of great white sharks grow up to two-and-a-half inches, and they have serrated edges that help tear the meat from the bodies of their prey. Obviously, such an animal can easily overwhelm many common sea creatures.

THE VILLAINS AND THEIR UNDERSEA ARSENALS

THE VILLAINS AND THEIR UNDERSEA ARSENALS

But their sheer power does not mean that great whites are indiscriminate feeders. In fact, quite the opposite is true. Like other predators, great white sharks fit into a specific niche in nature's overall scheme. Over the first few years of their lives, great whites prefer to feed on rays and flatfish such as halibut, turbot, and sole. When they are about four or five years old, their diet shifts toward marine mammals.

Worldwide there are more than 350 species of sharks, but surprisingly, more than 80 percent of all full-grown sharks are smaller than the average adult human being. The largest one, the whale shark, is a docile filter feeder. It preys on tiny shrimplike animals called krill and on other members of the plankton community by filtering large amounts of water through its mouth and out its gills. Whale sharks reach a size of just under 50 feet and 40,000 pounds, but they hardly fit the shark's typical image of a terrifying beast.

Upon first sighting, other species such as the horn shark and Port Jackson shark look more like catfish than sharks. Like angel sharks, wobbegongs, swell sharks, and other bottom dwellers, they blend into their surroundings on the bottom. Using camouflage as much as strength, these sharks feed on a variety of crabs, lobsters, sea urchins, snails, octopi, squid, and small fish.

Sharks can pose a serious threat to humans, and they should be respected. They will occasionally attack people, and in such a confrontation, the advantage always goes to the shark. But these animals fit into their own niches in nature, niches that do not include human beings as a primary food source.

Opposite page: *Despite its monstrous size, the whale shark won't hurt these divers. It eats mostly small ocean crustaceans filtered from the water.* Above: *This horn shark has eaten quite a few purple sea urchins in its time. The purple streak in its dorsal spine is a concentration of dye from the urchins.*

THE VILLAINS AND THEIR UNDERSEA ARSENALS

Above: *A lone barracuda is a common sight in the Caribbean, the Bahamas, and the Gulf of Mexico. They often cruise along reefs and over grass flats in their endless search for prey.* Below: *Apparently unafraid, a large barracuda silently watches this diver intrude into its world. Although they rarely attack humans, their persistence and large white teeth can be intimidating.* Opposite page: *Smaller barracuda often swim in schools.*

Barracuda are also potentially dangerous to swimmers and divers. These fish eaters have a mouthful of long, sharp, canine teeth that help them snag their fast-moving prey. Barracuda commonly reach six feet in length, and their tangled mass of teeth are almost always in full view; they are an intimidating sight. In addition, they have a habit of following swimmers and divers, often through an entire dive. While the experience can be unnerving, barracuda rarely attack humans, and when they do, carelessness almost always plays a part. Either spearfishers had their take tied to their bodies or a diver was tempting fate by hand-feeding a barracuda.

In natural settings, barracuda prey on small, silver-colored schooling fish such as silversides and mackerel. Occasionally, swimmers wearing jewelry do suffer bites from a barracuda. The flash of jewelry in the sunlight probably looks similar to the baitfish and fools the barracuda.

THE VILLAINS AND THEIR UNDERSEA ARSENALS

THE VILLAINS AND THEIR UNDERSEA ARSENALS

Killer whales, or orcas, are another of the marine predators whose capabilities and natural tendencies are often exaggerated and exploited. Most people who have dived with killer whales do not consider them a threat to humans. Like other species of toothed whales, killer whales use a form of natural sonar called echolocation. They emit sound waves that strike objects in the water and reflect back to them. The whales analyze the reflected waves to determine the objects' nature and distance. This and swimming speeds of 30 miles per hour help killer whales successfully prey upon salmon, sea lions, seals, dolphins, and even the largest creatures in the ocean, blue whales.

Above: *Killer whales have been intensely studied by scientists over the past 20 years, particularly in their northern feeding grounds. The whales seem to have become accustomed to human intrusion and are remarkably tolerant of onlookers.* Opposite page: *These whales swim in pods of about five to twenty animals. The pods usually consist of immediate family — a mother and all her offspring — and they remain together for their entire lives.*

How Killer Whales Got Their Name

The popular view of killer whales, or orcas, labels them as vicious marauders. In captivity, killer whales are wonderful animals to see, but people are generally frightened at the thought of encountering one in the wild.

They are big, and the towering dorsal fins of the males make them an imposing sight, but their reputation is really undeserved. Their English name is a misnomer and actually tells more about human behavior than whale behavior.

Several centuries ago, European and Russian hunters pursued orcas for their meat, skin, and oil in waters that Eskimos also frequented. In their own language, the Eskimos called these men "whale killers." When the term was translated to English, the words and the meaning became transposed, so that the animal received a new name and a bad reputation.

Another version of the story holds that Eskimos called orcas whale killers because they sometimes attack other whales. Again, in translation, the words became transposed, and orcas received an undeserved name.

Over the years, killer whales have paid a high price for the mistaken translation. As incredible as it seems today, after World War II the United States Navy considered killer whales a significant threat and actually made plans to bomb them. To date there has not been a single documented case of a killer whale attacking a person in the wilderness.

THE VILLAINS AND THEIR UNDERSEA ARSENALS

THE VILLAINS AND THEIR UNDERSEA ARSENALS

Perhaps the only marine mammal that poses a real danger to man is the leopard seal. Reaching a length of 12 feet, leopard seals live in the polar and subpolar waters of Antarctica where they readily hunt fish, penguins, and other seals. Leopard seals have numerous long, sharp teeth and a nasty temperament. They often lurk under holes in ice floes and suddenly rise up and grab any victim that has ventured too near the edge of the ice.

Moray eels are another of the more famous marine predators whose reputation belies their true nature. Morays are predators within reef communities, and they rely on their long, sharp teeth to snare prey. Unlike most fish, morays can swim both forward and backward, which helps them slither in and out of tight quarters as they attempt to outmaneuver their prey.

Morays do look mean. As they lie in the recesses of a reef, their jaws steadily work back and forth, as if they were gnashing their deadly teeth. It would be easy to misinterpret their eerie visage as a threat. Actually, the eels are just breathing. Morays are fish, but they lack the bony gill covers that other species use to draw water through their gills. Morays must continually open and close their mouths to pump water over their gills.

Morays are very wary of human intruders, and they often retreat into a nearby crevice upon sighting a diver. Actually their sense of sight is only fair. They rely heavily upon their well-developed olfactory system to locate their favorite prey of lobster, octopi, and fish.

Sea snakes are well known for their highly venomous toxins. Several species are far more poisonous than king cobras, but they are not aggressive toward humans. They are usually quite docile. Still, divers must keep the risks in mind.

Sea snakes are air breathers and need to be able to kill or immobilize their prey quickly. Pursuing and biting their standard diet of fish requires a lot of energy from the snakes. If a sea snake cannot swallow its captured prey whole, it must release its victim and go to the surface for air. If the neurotoxins were not fast acting, the prey would often escape. A fast and powerful venom enables the snake to return to the prey after going to the surface for air.

Opposite page: *Sea snakes are highly venomous. While they are not normally aggressive toward humans, prudent divers will always avoid them in the wild. They live only in the tropical waters of the Pacific and Indian Oceans.* Above: *Moray eels appear to be a much greater threat than sea snakes. In reality, however, they will only bite if provoked or tempted with food, and their bite carries no venom.*

THE VILLAINS AND THEIR UNDERSEA ARSENALS

This page, top: *Floating on the water's surface at the mercy of the wind and the currents, a seemingly helpless Portuguese man-of-war trails tentacles loaded with powerful stinging nematocysts. Few fish that stumble into the tentacles ever come out.* Above: *A stingray's tail holds barbs that can be deftly aimed at any potential predators.* Opposite page: *A relative of the shark, stingrays spend most of their time lying on the seafloor. Swimmers familiar with their habits have learned to shuffle their feet when they walk through shallow areas to avoid surprising the rays and prompting them to strike.*

Some of the most dangerous sea creatures use toxins injected by stinging cells to capture their prey. Perhaps the best-known stinger is the Portuguese man-of-war. Often thought to be jellyfish, man-of-wars are actually hydroids closely related to the jellyfish. At the water's surface, they appear as a floating purple-to-blue iridescent balloon. Below the balloon, or float, the man-of-war trails a mass of stinging tentacles that can stretch down 100 feet, although most only reach about 20 feet. Each tentacle wields thousands of stinging cells containing potent toxins that disable their prey within seconds.

Portuguese man-of-wars feed primarily on fish. The dangling tentacles lure curious fish into contact with the stinging cells and their harpoonlike barbs. The man-of-war then pulls the tentacles and food up toward its mouth.

Many people think stingrays also defend themselves by stinging. Actually the stingrays' tails carry long barbs that are more like jagged knives than stingers. Stingrays possess incredibly flexible bodies, and they can aim the barbs with deadly accuracy. They often lie in shallow sandy areas, and the barbs keep other animals from troubling them while they rest in the sand. They have also taught people to step cautiously when they're wading.

Lionfish, stonefish, and scorpionfish are bottom-dwelling reef fish that use spines as a means of defense. All have toxic spines in the dorsal fins atop their backs, and some species have venomous spines along their sides as well. All of these fish are slow

THE VILLAINS AND THEIR UNDERSEA ARSENALS

THE VILLAINS AND THEIR UNDERSEA ARSENALS

swimmers, but there is an interesting contrast in their outward appearance. A favorite in marine aquariums, lionfish possess long, ornate fins and spines, and many specimens have bright color patterns. As a result, they stand out in their natural setting. On the other hand, scorpionfish and stonefish are well camouflaged.

The contrast in appearance shows that in nature there is more than one way to solve a problem. In the case of lionfish, their "obviousness by design" advertises their potent spines. By being so easy to see, they hope that other animals will recognize them as a danger and leave them alone. Well-camouflaged scorpionfish and stonefish take a different approach. Their first line of defense is hiding, and their toxic spines serve as a backup protection.

Certainly the animals mentioned in this section are all impressive hunters. They must be in order for their species to survive. But just because they excel in their natural roles does not mean that they pose a significant threat to humans. While it makes sense to stay out of the water if great whites or man-of-wars are around, the odds of an attack by any sea creature are extremely low.

SOME LESS FAMOUS VILLAINS

The legends of sea lore—sharks, eels, killer whales, barracuda, and all the rest—come to mind quickly when the topic is dangerous marine life. These beasts are not the only ones that must capture their food and defend themselves, however. All creatures, large and small, must be able to perform these tasks. The adaptations of the large, dramatic predators are infamous, but many other, lesser-known marine animals are every bit as deadly. While these animals get far less attention than animals that have a mouth full of big, sharp teeth, many pose a far larger threat to humans.

To survive through the eons, those species lacking the obvious advantage of brute strength have evolved other means of coping. For example, bristle worms, or fire worms, are colorful, delicate animals that live in coral reef communities. They are slow crawlers and even worse swimmers, and their ornate design makes them stand out against their surroundings. Like all other animals, they must protect themselves from fish and other predators. Bristle worms bear tufts on their sides that house thousands of long, sharp, fiberglass-like spines, or bristles. The spines can easily penetrate human skin and cause intense pain that lasts for hours or even days.

Opposite page: *The bright stripes and long fins of the lionfish advertise the potent toxin it carries in its dorsal spines.* Above: *Certainly less beautiful than the lionfish but no less deadly, a scorpionfish hides in a rocky outcrop. Scorpionfish also wield toxin in their dorsal spines, but they prefer to disguise themselves, skillfully blending in with their surroundings.*

THE VILLAINS AND THEIR UNDERSEA ARSENALS

Delicate-looking coral polyps also have potent stinging cells. Corals are animals that permanently attach themselves to the sea bottom. They cannot pursue their prey, nor can they move to avoid predation. It is critical that their stinging cells be both powerful and fast acting. If anything from a microscopic organism to a human being brushes against a coral, the coral will strike. The stinging cells of some species are too short to penetrate human skin, but others inflict sharp, painful stings that can be very slow to heal.

Above: *Even some small invertebrates like the bristle worm have formidable, spiny defenses. Each bristle is hollow and filled with venom.* Opposite page: *Batteries of stinging cells cover the tentacles of each soft coral polyp. The stings are used more for feeding than for defense, but they do serve as an effective deterrent.*

Colorful sea anemones, hydroids, and jellyfish are other animals whose delicate appearance masks a potent sting. Often these creatures are difficult to see and avoid because they are small and semitransparent. Though it is not common, some jellyfish stings can actually be fatal to humans, especially small children. It is, however, common for the stings of all of these animals to kill their intended prey.

Formidable spines and a hard shell protect the soft bodies of many species of sea urchins. While the spines are obvious, their clever design is not. They penetrate the body of any animal that applies pressure to the urchin, but they cannot easily be removed since the spines are brittle. Efforts to remove them often leave the tips buried just below the skin. Humans normally eject the spine only after the area becomes infected. It is a lesson not quickly forgotten. In natural settings, the spines work just as effectively against most enemies.

Cone snails are among the most attractive and most dangerous of all invertebrates. A variety of stunning patterns decorate their shells, making them a beachcomber's dream. However, their proboscis, a long mouth-like organ, contains an extremely toxic tooth. The proboscis is long and flexible, and can whip around and strike an animal that is behind the snail. Often, even experienced divers do not recognize cone snails as dangers. Some will pocket live cone snails while on a dive only to receive a fatal sting when they remove their prize back on shore.

THE VILLAINS AND THEIR UNDERSEA ARSENALS

THE VILLAINS AND THEIR UNDERSEA ARSENALS

THE VILLAINS AND THEIR UNDERSEA ARSENALS

Another potentially dangerous yet very attractive denizen of the deep is the blue-ringed octopus of South Australia. While their cousins the giant octopus and giant squid get far more publicity, these softball-sized octopi are a greater threat to humans. The octopus is not at all aggressive. In fact, it is usually rather shy. However, like all octopi, blue-ringed octopi possess strong parrotlike beaks, and they can inject a powerful neurotoxin into their prey. If handled, they sometimes bite.

While large, dramatic creatures such as great white sharks and killer whales receive far more media attention, a multitude of other species employ formidable, albeit discreet, defense tactics. In many instances, their adaptations are less obvious, at least to humans, but they serve their users well.

Opposite page: *A crown of tentacles tops the soft body column of this anemone. These animals feed like a single, large coral polyp, waving numerous stinging tentacles in the water and waiting for passing prey.* Above: *The blue rings on this octopus are a sure sign that it's poisonous.* Below: *Long spines keep most predators from attacking this spiny sea urchin.*

INVERTEBRATES

The kingdom of invertebrates is a broad one. In the marine world, it covers animals as different as the tiny coral and the enormous giant squid, or the primitive sponge and the highly developed octopus.

An invertebrate is any animal lacking a backbone. Other than that, members of the group have very little in common. Some have sophisticated internal structures, and others are little more than a loosely joined mass of cells. Sea stars normally live their lives in isolation, while barnacles settle in large groups and bryozoans form colonies where the individuals tailor their body structure to benefit the group.

Invertebrates are the most varied and numerous life forms in the sea. Representatives abound in every ocean environment, and they do much to shape their habitats. They serve as the primary food source for many higher animals, but they do much more. Some corals literally shape reef communities. Sponges feed on debris that might otherwise cloud the water and alter living conditions. And as a group, invertebrates add much to the beauty and wonder of the marine world.

The invertebrates are a broad group, containing animals with a wide variety of body shapes, lifestyles, habitats, and histories. Opposite page: *The ten arms of this sea star radiate from the central body, or disk, of the animal. Although most species of sea stars have only five arms, others can have many more.* Above, left: *The eight arms of an octopus, each studded with suckers used for grasping prey and for moving, coil and uncoil as the mollusk moves almost effortlessly along the sea floor.* Above: *The many tentacles of the club-tipped anemone wave in the water waiting for passing prey. Any food they catch is passed to the slit-shaped mouth in the center of the tentacle ring.*

INVERTEBRATES

SPONGES

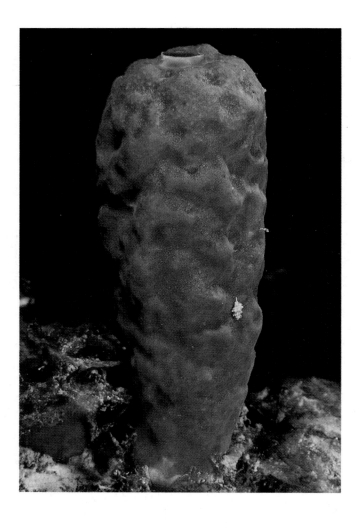

This page, top: *One distinguishing feature of this purple sponge is the single round pore, or osculum, at its top. Water is drawn into the sponge through countless pores all over its surface and filtered for food and oxygen, then it exits through the osculum.* Above: *This tangle of red rope sponge has several oscula along each of its long strands.* Opposite page: *Imposing columns of this yellow tube sponge jut up from the bottom. Sponges come in all the colors of the rainbow.*

Without muscles, nerves, and even specialized tissues, sponges are one of the most primitive multicellular animals. They consist of several types of loosely gathered cells. Worldwide there are about 5,000 species of marine sponges, and another 150 varieties live in fresh water.

In nature, most adult sponges attach themselves to corals, rocks, debris, or the shells of other animals such as hermit crabs and decorator crabs. Sponges occur in an incredibly wide variety of colors and shapes. Many simply take on the shape of whatever they happen to be growing on. Some sponges are symmetrical and others irregular. Some sponges are low and flat, while others are erect and occur in lobes. Most are colonial, but some are solitary. Although adult sponges attach themselves to the bottom, they are free swimming in their larval stage.

In some settings, sponges are among the most conspicuous members of their habitat. Found in the waters off Belize on Turneffe Reef and in many other parts of the Caribbean, bright yellow tube sponges grow in clusters and reach a height of more than three feet. Stretching up more than six feet, barrel sponges dominate many tropical reef seascapes. Brilliantly colored red and orange cup sponges, vase sponges, glass sponges, and basket sponges are other examples of large, rather obvious species. Yet other species can be small, mostly hidden, and drab. These sponges will often go almost unnoticed by all but dedicated specialists.

INVERTEBRATES

INVERTEBRATES

Porifera, the scientific name given to the phylum of sponges, refers to the numerous pores, channels, and openings in their bodies. These canals form a system that allows water to flow through sponges. The water brings life-sustaining food and oxygen and also removes waste products.

Some species of sponges are borers, eroding the limestone structure of reefs and destroying the shells of many mollusks. Most species, however, are filter feeders. Sponges play a major role in the ecology of many reef systems by filtering the water. As water passes through their channels, the sponges consume bacteria, debris, and other small floating particles. Eighty percent of the organic matter sponges ingest is too small to see with an ordinary microscope.

Sponges have few natural enemies. Their foul taste and noxious secretions make them painfully unappetizing to most animals. Still, some species of angelfish, filefish, and sea slugs do prey upon sponges. And other fish, brittle stars, worms, shrimps, crabs, juvenile lobsters, and snails use the hidden recesses of sponges as a place to hide.

Some, but not all, sponges have a commercial value. Common bath and household sponges are actually the skeletons of sponges that have had the living tissue removed. Today, many household and industrial sponges are synthetic. Some species are also used to produce antibiotics and other medicines.

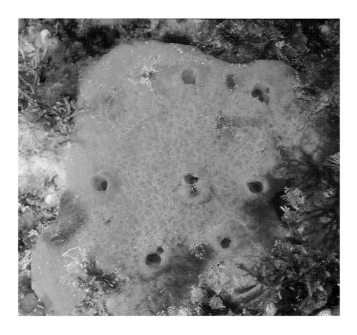

Opposite page: *The smokey cloud rising from this large basket sponge is composed of sex cells. Sponges spawn by releasing male or female gametes through their osculum with the exiting water current.* This page, top: *Not all sponges are large and impressive. Many grow in thin crusts over any hard surface that they can gain a foothold on.* Above: *The long, white antennae protruding from the top of this sponge are a clear sign that some coral-banded shrimp have taken up residence.*

INVERTEBRATES

THE STINGERS

Coral polyps and sea anemones are often called the "flowers of the sea" because their flowing tentacles resemble flower petals. But make no mistake, the similarity ends there. Corals and sea anemones bear stinging cells, primarily in their tentacles. These creatures are close relatives of jellyfish and Portuguese man-of-wars, animals that even most nonspecialists recognize as potentially harmful. All of these creatures, along with sea wasps, sea pens, sea pansies, and hydroids, are members of the phylum Cnidaria. Their most distinguishing characteristic is their stinging cells, called cnidocytes.

Inside each stinging cell is a structure called a nematocyst. Although nematocysts vary from one species to another, most look somewhat like coiled harpoons. When stimulated chemically by potential prey and predators or physically by touch or increased water pressure from even slight movements, the nematocysts fire. The time between initial contact and the stinging cells' response is a bare fraction of a second, making this one of the fastest cellular responses in nature.

As a rule cnidarians are very poor swimmers and crawlers; they can't stalk or chase down their prey. When they stumble across a meal, they need to secure it quickly. Released by the nematocysts, their powerful, deadly toxins can induce almost immediate paralysis in small animals. In some cases, human skin is too thick for the nematocysts to penetrate. In others the nematocysts can inject their toxins and cause severe itching, swelling, and even death.

Above: *In the Caribbean, the wide, lacy oral arms and thin tentacles trail below the umbrella-shaped bell of a jellyfish medusa. Both oral arms and tentacles bear stinging cells.*
Opposite page: *On the other side of the world in the South Pacific, a red sea fan waves back and forth in the ocean currents, trapping passing prey and soaking up the tropical sunlight with its branched tentacles. The black and white feather star perched atop the sea fan is another invertebrate that strains plankton from the water.*

All cnidarians occur in two basic forms. Adult corals, sea anemones, and hydroids grow as polyps. Polyps are attached to the sea bottom by a footlike disk. The mouth is at the opposite end and faces into the water. The second form, that displayed by adult jellyfish, is called a medusa. In this form the animal is free swimming and the mouth points downward. The life cycle of many cnidarians includes both phases—an asexual, bottom-dwelling polyp stage and a free-swimming, sexual medusa stage.

INVERTEBRATES

INVERTEBRATES

Corals

In their adult stage, which is the way most people think of them, corals occur in a wide variety of forms. Corals appear in three major categories: the hard or stony corals, many of which form reefs; the thorny corals such as the black coral; and soft corals like the elegant sea fan.

All true reef-building corals are colonial. The colonies live atop the limestone skeletons of their ancestors and form the world's tropical reefs, making them one of the most significant invertebrate animals in warm, shallow seas. Without the reef-building corals, the topography of tropical oceans would be very different. For reef-building corals to flourish, the water temperature must be above 70°F, and there needs to be some water movement.

Most hard corals are sedentary and depend on their tentacles, armed with stinging cells, to capture plankton that drifts by in currents. When feeding, the polyps appear flowerlike as they open up and extend into the water column, waiting to sting any unsuspecting prey. When not feeding or when threatened, the tentacles retract, creating an entirely different outward appearance.

The common names of the hard corals demonstrate the seemingly endless variety of shapes and patterns of the colonies. Elkhorn, staghorn, star, brain, pillar, and lettuce corals are some of the more important hard corals in the Caribbean. Tabletop, finger, flower, mushroom, and bubble are the common names of other species.

Opposite page: *Despite its bright color, this rosy branching coral can be hard to spot because of its small size and its preference for dark caves or overhangs. Although no two colonies of the same coral are ever identical, the general shape and often the color are characteristic of the different corals.* Top: *Elkhorn corals produce large branching structures that contribute to the growth of the reef itself.* Above: *Some hard corals, such as the fungus corals, form flat disk-shaped colonies.*

INVERTEBRATES

INVERTEBRATES

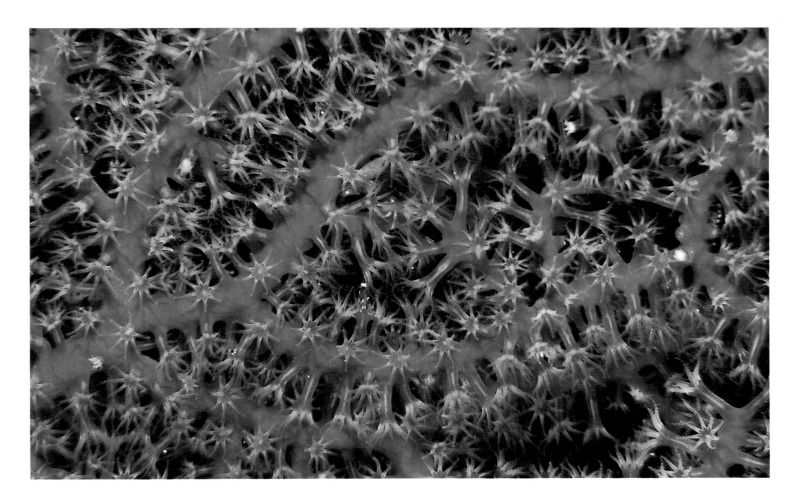

Thorny corals secrete a tough skeleton made of proteins. The skeletons form a large number of outer spines, which is where thorny corals get their name. These corals live only in warm seas, usually at greater depths. Black coral is one well-known thorny coral. Underwater black coral trees actually look yellow. But beneath the living yellow polyps lies a black skeleton that after being polished is used as jewelry. Not too many years ago, black coral was not uncommon. However, it has suffered commercial exploitation for several decades and is now much harder to find.

Soft corals, unlike hard corals, are not reef builders. Their bodies do not develop hard skeletons as do the bodies of hard corals. To the eye of the underwater photographer, however, few subjects offer as much potential.

Soft corals live in tropical and temperate seas around the world, and they, too, occur in many different forms. Soft corals include the animals commonly called sea fans or gorgonian corals, sea whips, and wire corals.

Dazzling arrays of brightly colored soft corals provide a rare visual treat for divers. The colors of soft coral trees range from cherry red to burgundy to sunshine yellow to vivid orange to snow white. Colonies can stand taller than a person, and in areas where current flow is steady, a diver will discover tree after tree along a coral reef for the duration of a dive.

Opposite page: *Black corals form pliable, finely branched colonies. They are most often found in deeper water growing out from vertical walls.* Above: *The individual white polyps give these red Philippine gorgonian corals a soft, fuzzy appearance. If the colony is disturbed the polyps will retract.*

INVERTEBRATES

74

What Are Zooxanthellae?

One correct way of answering the question "what are zooxanthellae (pronounced zoh-zan-thel-lee)" is to say that they are microscopic single-celled algae. However, a more revealing answer is that zooxanthellae are an integral part of coral reef communities and the solution to a mystery that puzzled scientists for years.

Almost all major food chains on earth begin with plants, which convert energy from the sun into simple sugars through photosynthesis. Within food chains, plants are producers. Animals that derive their energy by feeding directly on the producer plants are called first-level consumers. Other animals that feed on the first-level consumers are called second-level consumers, and so on. However, as energy moves up each level of a food chain, almost 90 percent of it dissipates. In other words, a first-level consumer needs to eat 10 pounds of plants to gain one pound. A second-level consumer needs to eat 10 pounds of first-level consumers, the equivalent of 100 pounds of plants, to gain one pound.

For years scientists really did not understand how coral reef communities worked. Although some large plants are present in most tropical reefs, they are not abundant. At first glance, it appears that there are far more consumers, such as fish and invertebrates, than there are producers.

The answer lies in huge quantities of very small plants. In shallow waters, algae lives at the sea bottom and on dead corals. Algae also lives inside the polyp tissues and skeletons of the corals themselves. Zooxanthellae are the algae living inside the corals, often in incredibly large numbers. These tiny plants work as the producers in reef communities, serving as the foundation of the reefs' food chains.

The zooxanthellae share a vital relationship called mutualism with the coral polyps. The algae assist the coral in its ability to grow and build a skeleton by providing raw materials through photosynthesis. In turn, the zooxanthellae use the waste products given off by the corals.

Opposite page: *Bright yellow tentacles adorn a long, thin wire coral off Fiji. These corals grow in single, long, unbranched stalks that can be several feet long.* Left: *Off Belau, pink sea whips grow in tufts that branch near the base.*

INVERTEBRATES

INVERTEBRATES

A Hard Coral Polyp

The polyps of hard corals usually appear quite small to divers and snorkelers. The polyps secrete materials to form the small cups that they live in. At their center is an opening surrounded by tentacles that serves as both mouth and anus. Below the opening lies the gastrovascular cavity where food is digested.

Inside each tentacle are a number of stinging cells called cnidocytes. And inside each cnidocyte is a structure called a nematocyst. The nematocyst contains a small threadlike filament. When prompted by chemical stimuli or by touch, the nematocyst releases the thread. There are at least 30 different kinds of nematocysts, and many are capable of delivering some type of potent toxin.

Some corals feed by brushing their tentacles across their mouth after capturing prey. Others use a system of cilia and mucus to gather their food. The polyp spreads a self-made mucus trap. After the prey is trapped, the hairlike cilia pull the trap toward the mouth.

A thin horizontal sheet of tissue joins all the polyps of a coral colony. There is some evidence that the responses of some colonial polyps are somehow interconnected. Contact with any one polyp can make nearby polyps retract.

Opposite page: *The typical coral polyp is tube shaped, with a ring of tentacles surrounding a mouth at the top, like these from the Red Sea.* Top: *Orange flame corals thrive in the Caribbean. Many polyps, connected by living tissue, make up each coral colony.* Above: *Coral polyps can contract down into individual, cup-shaped depressions in the colony skeleton so that the tentacle tips are barely visible. With the tentacles retracted, the corals pose no threat to this small fish.*

INVERTEBRATES

78

Sea Anemones

Sea anemones are similar to their close cousins the corals, but anemones have much larger polyps. Both look like a ring of tentacles connected to a circular disk. Some anemones prefer a solitary existence, while others live in clusters. When an anemone is feeding, its tentacles flow gently back and forth with the movement of the water, providing a visual treat for divers.

In temperate waters along the coast of western North America, baseball-sized green solitary anemones thrive. In tropical waters, some anemones can grow to more than three feet in diameter. Some anemones, especially those that live in the sand, reside in tubes made of mucus and debris.

Opposite page: *Like corals, sea anemones are simple feeding machines. The single opening in the center of the tentacles is the mouth.* Above: *Many anemones feed on plankton, but larger species will readily consume small fish that wander too close.* Below: *Although anemones come in a variety of shapes, sizes, and colors, they all have the same basic structure.*

INVERTEBRATES

Strange Bedfellows

Anemonefish, often called clownfish due to their bright colors and amusing antics, enjoy an unusual relationship with sea anemones. Found throughout the Indo-Pacific and Red Sea, clownfish are able to safely swim among the deadly tentacles of sea anemones without getting stung. Armed with a multitude of stinging cells in their tentacles, anemones capture their food by stinging their prey and pulling it toward their mouth with their tentacles. While most of the food anemones eat are tiny organisms, some can capture fish. However, all clownfish are able to spend hours on end inside an anemone's mass of tentacles.

The clownfish benefit from this in several ways. First, the anemone offers a safe place to hide. Potential predators cannot pursue the anemonefish without almost certainly suffering fatal stings. In addition, when the anemone captures food the clownfish may be able to take some of the scraps for themselves. In some cases, the clownfish feed upon waste products from the anemones.

The anemone benefits because the fish help remove parasites and waste products. The anemonefish may also drive away some potential predators of the anemone. It is also possible that the anemone feeds on scraps from the clownfish's meals.

Exactly how clownfish avoid being harmed by the anemone is an interesting story in itself. Though the whole process is still unclear, scientists do understand parts of it. The fish begin with very brief contacts with the anemone. Despite the stings, the contact continues and gradually intensifies. Eventually the anemonefish are able to rest in the anemone's tentacles without any harm. The fish produce a protective layer of mucus that surrounds them. If a fish leaves the anemone for an extended time, it must reacclimate to that particular anemone.

Right: *Most small fish would be eaten by an anemone this size, but clownfish have adapted themselves to life within the dangerous stinging tentacles.* Opposite page: *Also called anemonefish because of their interesting association with these invertebrates, this group of orange fish are all marked with one or more white vertical bars.*

INVERTEBRATES

INVERTEBRATES

Jellyfish

Jellyfish are relatively poor swimmers. While they are able to control their attitude and direction somewhat by contracting their bell, they usually go wherever currents and winds push them. Because they can't actively hunt their prey, they must rely on unwary or curious animals coming into contact with their tentacles. In many species, the tentacles are semitransparent and very difficult to see. The tentacles trail underneath the bell of the jellyfish, waiting to ensnare a victim. Although many jellyfish appear quite small, their deadly tentacles can extend down 100 feet.

As a result of their limited mobility, jellyfish will never actually attack a swimmer. On the other hand, because the currents and winds move them, they are often found in concentrations. If a swimmer encounters one jellyfish, there are likely to be a lot more in the area.

Most jellyfish can inflict stings that range from barely noticeable to extremely painful and even fatal in humans. It is important to know that the firing of the stinging cells is a mechanical process. Even after a jellyfish is dead and has washed up onto a beach, unfired nematocysts may still be able to unleash their toxins when stimulated. Not all jellyfish stings harm humans. The moon jellies that live in the Caribbean have very short tentacles, and most people feel only a slight prickly sensation if stung.

This page, top: *This comb jelly has captured a small crustacean, which will quickly be digested.* Above: *The moon jellyfish, with its scalloped bell margin and short fringe of tentacles, can grow larger than a dinner plate.* Opposite page: *The winds and currents often concentrate jellyfish in swarms, which can be a serious problem for swimmers because the jellies pack a powerful sting.*

INVERTEBRATES

INVERTEBRATES

84

Portuguese Man-of-wars

Surprising to many laymen, Portuguese man-of-wars are not true jellyfish. They are closer relatives to hydroids than to jellyfish because of similarities in their body structure. Man-of-wars are actually colonies of individual animals in both the medusa and polyp stage.

Portuguese man-of-wars are visible as they float on the surface. Their iridescent blue to purple bells, or floats, look like irregular, wrinkled sacs made of plastic wrap. Below and behind the bell trail a number of very potent tentacles that can easily kill prey such as mackerel and anchovies.

Hydroids

Many seasoned divers do not even know hydroids exist because many of them are so inconspicuous. However, anyone who has come in contact with their stinging tentacles is not likely to forget them. Most hydroids are colonial and live in arrangements that look a lot like feathers. Some are solitary. The solitary species are quite small, but their stings can be intense. Hydroids often attach to the end of sea fans and sea whips. From this vantage point, their tentacles can reach into the water column and snare drifting plankton.

The by-the-wind sailor, often mistaken for a jellyfish, appears in clusters at the surface. The float of a by-the-wind sailor is a flattened sheet that looks like the sail of a sailboat and functions in much the same way. The float breaks the surface and uses the wind to create thrust.

Opposite page: *A small part of the Portuguese man-of-war floats on the water's surface, carried by currents and blown along by the wind. Its tentacles, however, can trail up to 100 feet below.* This page, top: *By-the-wind sailors are also buoyed up by a gas-filled float. The float is topped by a thin sail from which these jellies get their common name.* Above: *Feathery, branched hydroid colonies are closely related to jellyfish. One of the things they have in common is their stinging cells.*

This page, top: *Sea pens may be the strangest of the cnidarians, but like the corals, anemones, jellyfish, and hydroids, each pen is decorated with numerous little polyps.* Above: *When the polyps are retracted, the sea pen looks quite different.* Opposite page: *Brightly colored flatworms should be easy to spot creeping along the bottom, but many are small and hide in crevices and under rocks.*

Sea Pansies and Sea Pens

Colonial animals, sea pansies are flat, disk-shaped creatures anchored in the sand by a stalk. The polyps work together to form a large mucous net that traps prey. When the net captures any prey, the sea pansy consumes both the net and the ensnared victim. Many sea pansies are bioluminescent, meaning they create their own light. At night, divers can make a sea pansy glow by gently stroking it.

Sea pens are another form of colonial cnidarian that live in the sand and create their own light. The fleshy sea pens of the temperate seas along the coasts of New England and the western North America are among the most beautiful and delicate of all sea creatures. Sea pens have stalks. When feeding, the polyps extend and give the colony a soft, bushy appearance. However, when threatened the polyps retract, and the sea pen looks similar to a dead leaf. In some species, the stalk will retract into the sand or mud to avoid a threat.

FLATWORMS

Scientists distinguish between many different types of worms. Each appears in its own phylum. Flatworms are exactly that — flat. Many free-living marine flatworms are also quite colorful. Lay people often confuse the more brilliantly colored species with nudibranchs. With just a little knowledge, it is relatively easy to tell the two groups apart. The bodies of flatworms are very thin, much more so than the nudibranchs.

INVERTEBRATES

INVERTEBRATES

Most of the 3,000 or so species of marine flatworms are less than one inch long. However, several tropical flatworms exceed six inches in length. When disturbed, the free-living worms swim in a series of rhythmic undulations that many divers find captivating.

Being closely related to terrestrial tapeworms, most marine flatworms are bottom dwellers, and some species are parasitic. Flatworms have a rather simple physiology, but in a biological sense they are very significant. They are the most primitive animals to display bilateral, rather than radial, symmetry. This means that they have a distinct left side and right side that are mirror images of each other. Other advanced features are a concentration of nerve cells in their front end, or head, and a distinct excretory system. While flatworms have few specialized organs, many possess clusters of light-sensitive cells that are often quite prominent.

Many flatworms are cannibalistic, and many reproduce sexually, a combination that presents a rather interesting problem. During the breeding season, individuals can identify potential mates that will not eat them. If that were not the case, mating could lead to the demise rather than the growth of the species.

SEGMENTED WORMS

In the beauty contest of the world of worms, there is little doubt that the segmented marine worms put their terrestrial cousins to shame. Surprising as it sounds, worms are among the most beautiful and delicate of all marine creatures. Many new divers have trouble believing that these exquisite creatures are lowly worms.

In their natural habitat, many segmented worms don't look at all like the terrestrial worms that people know and loath. Fan worms, feather dusters, plume worms, and Christmas tree worms look more like colorful fans, feathery flowers, or miniature Christmas trees than they do worms. However, divers do not normally see the entire

Above: *Like most invertebrate groups, flatworms come in many colors and sizes. This bright yellow species can be found in the waters off New Guinea.* Opposite page: *The whorled gills of a Christmas tree worm extend into the water for respiration and for filtering out small food particles.*

worm. The exposed part of the worm is actually a group of branched tentacles called radioles that trap plankton from the passing water. The rest of the worm, which is usually the vast majority of the animal, lives burrowed in the coral, rock, mud, or sand, or in a self-made tube. If removed from the bottom or from their tube, the bodies of these marine worms look very similar to their terrestrial relatives.

In terms of their physiology, the segmented worms are far more advanced than marine flatworms. Many annelid worms possess specialized sensory organs concentrated mainly near their heads. The annelids have a more advanced nervous system than the flatworms, and a few species' eyes are not only light sensitive but also have lenses.

Segmented worms probably evolved in the sea. As the name suggests, they have clearly marked segmented bodies. Most annelids have a pair of paddlelike appendages on each segment of the body used in swimming or moving across the bottom. Projecting from these appendages are many bristles, or setae. Though very thin, these bristles are long, sharp, and sometimes poisonous. The bristles form a formidable wall around the worms and serve them well. The bristles function for defense or to assist in moving.

The most noteworthy of the segmented worms are the fire worms or bristle worms. Both common names are appropriate. If touched, even ever so slightly, their fine bristles will pierce human skin and often cause severe pain and swelling. Marked with vibrant, colored segments bordered by white setae, bristle worms appear harmless, but even gloves don't always provide protection against their sting.

Equipped with powerful jaws and fangs, bloodworms are voracious predators. They build a series of tunnels in the muddy bottom. When they detect motion near a tunnel's opening, the worms quickly seize and devour their prey, mainly crustaceans and other invertebrates.

This page: *The feather duster worm will retract its frilled gills into a self-made tube at the slightest provocation. However, a patient observer will be rewarded as these cautious animals poke back out.* Right: *Spiral gill worms build tubes down into hard coral colonies. A single coral may be peppered with many worms.*

INVERTEBRATES

91

INVERTEBRATES

BRYOZOANS

Inconspicuous by size, bryozoans are remarkably beautiful colonial animals. Many experienced divers know very little about them. They see bryozoans on reefs and refer to them as "delicate, pretty, lacy-looking...things." But once divers, especially underwater photographers, have seen them, they watch for them because of their visual appeal.

Bryozoan colonies take on a variety of shapes. Some arrange themselves in flat sheets, while others grow in upright lobes, plantlike tufts, or lacy growths. Bryozoans attach themselves to sand, rocks, kelp, coral skeletons, and other animals. The colonies range in color from drab brown to snow white to soft pinks and handsome tans. Many colonies are fairly small, only a few inches across, but some colonies can be several feet wide.

The word bryozoan is derived from two Greek words: "bryo" means moss, and "zoan" means animal. In some respects bryozoans look quite a bit like small patches of moss. There are about 4,000 species of bryozoans around the world, but scientists know little about the animals as a group.

The individual animals, called zooids, are often so small that it takes a microscope to distinguish between individuals. Each animal consists of an elongated body topped by a ring of tentacles that surround the mouth. They lack specialized organs for respiration, circulation, and excretion, but within a colony, various individuals may be specialized in their roles. For example, some individuals are equipped with a tiny jaw to discourage intruders.

Most bryozoans are hermaphrodites, with each individual being both male and female at the same time. In some cases their gametes, or sex cells, are produced alternately, and at other times the same animal simultaneously produces sperm and eggs.

Opposite page: *Bryozoans, also called moss animals, will grow in patches on almost any underwater surface. Colonies can be branched and erect, or encrusting.* Above: *Some bryozoans look like intricately tatted lace in miniature.*

INVERTEBRATES

MOLLUSKS

The phylum of mollusks contains such seemingly diverse animals as snails, octopi, scallops, and nudibranchs. As unlikely as it seems, evolutionary specialists believe that all mollusks arose from a single group of common ancestors. Today approximately 50,000 species of mollusks inhabit planet earth, and representatives are found in every marine habitat. Mollusks are well-developed animals, having distinct organ systems and senses. In fact, many experts believe that octopi are the most intelligent marine invertebrates.

Mollusks share a number of common characteristics, many of which are unique to members of this phylum. All have a soft, fleshy body called the visceral mass; all have a mantle that usually secretes a shell; and all have some type of muscular foot that is used in locomotion or in digging. Snails use their foot to walk or crawl, while clams dig with their modified foot. One other common feature is that the sexes are usually separate.

Many, but not all, mollusks have an external shell. The shell is obvious in snails, oysters, and rock scallops, but not as evident in squid and sea hares. Most of the animals in the latter group have a vestigial shell that they carry internally. Others, such as octopi and nudibranchs, have completely lost their shell.

Opposite page: *Octopi are only one of the many different types of animals that form the mollusk family.* Top: *Although a nudibranch bears no resemblance to the octopus, it too is a mollusk.* Above: *The most familiar of the mollusks are probably those with shells, such as this horse conch.*

INVERTEBRATES

Snails, Abalones, Conchs, and Cowries

The largest and most diverse class of mollusks, the gastropods, or stomach-foots, include the snail, abalone, conch, and cowrie. In one form or another, gastropods are found in all oceans.

Snail shells occur in a wide variety of patterns and shapes, although the color and structure is consistent for any given species. Many snail shells have a conical spire shape with a series of whorls. In some snails, beautiful, intricate patterns cover the shell's surface.

In most species, the animal remains completely hidden within the shell. However, if divers cautiously approach a seemingly empty shell, they often catch a view of the snail's stalked eyes, tentacles, and mouth. A snail's eyes rest on the ends of long, movable stalks, allowing the slow-moving snail to scout its surroundings without exposing its soft, vulnerable body.

Snails move by secreting two kinds of mucus. One is a sticky, gellike mucus that anchors part of the foot. The other is soft and slippery; it eases friction and lets the moving part of the foot glide over rough surfaces.

Some gastropods are an excellent food source for humans. Along the west coast of the United States and Canada, the favorite culinary species include abalone. Abalones look like flat, round rocks with a series of holes and several short tentacles. The meat comes from the abalone's large, muscular foot. The strong foot enables the abalone

Opposite page: *The coiled shell of a moon snail in the Solomon Islands is girdled in white trim by the animal's foot. When the snail is disturbed it can retract inside its mobile house.* Above: *Much to the disappointment of many shell collectors, the colorful rings on the flamingo tongue cowrie are part of the animal, not the shell.*

to firmly attach to rocks. Abalones' major predators include bat rays and sea otters.

Many snails graze on algae, using a rasping tonguelike organ called a radula. Some snails, like the horse conch, are active predators, mainly devouring other mollusks.

Among the most active carnivorous snails is a particularly attractive and highly venomous group called cone shells. Found in the tropical and subtropical waters of the Indo-Pacific and the Atlantic, these snails inject potent toxins into their prey with a detachable radular tooth. Because their shells are so beautiful, some people like to collect cone snails. But several Indo-Pacifc species can be fatal to humans. All too often, inexperienced collectors pick up a cone snail during a dive and put the animal in their pocket. The cone shell then injects its fatal poison.

In most snails, the sexes are separate, but as with many invertebrate groups, there are numerous exceptions. One of the most fascinating is the common slipper snail, *Crepidula fornicata*. These snails usually live permanently attached to one another in stacks. All begin their lives as males and later some turn into females.

INVERTEBRATES

Nudibranchs and Sea Hares

Nudibranchs are mollusks without shells that are famous for their dazzling colors and unusual shapes. Their name derives from Latin terms meaning "naked gills." Nudibranchs lack the external shell found in most mollusks, which leaves their gills exposed.

Nudibranchs occur in a wide variety of sizes, in bizarre shapes, and in an almost endless combination of colors. Words alone simply do not do them justice. Ranging in length from less than ½ inch to nearly two feet, nudibranchs are important members of reef communities throughout tropical and temperate seas.

Some of the best-known species include the Spanish dancer, the Spanish shawl, and the lettuce slug. Approaching 24 inches, Spanish dancers are among the largest species. Their bodies are usually a shade of orange or bright red, and some specimens have a border of yellow or white. Spanish dancers obtain their name from their practice of swimming above the reef in a captivating motion reminiscent of Spanish flamenco dancers. In California alone there are more than 100 species of nudibranchs, one of the prettiest being the purple and orange Spanish shawl. Lettuce slugs inhabit Caribbean reef communities.

Some species spend the majority of their time grazing on algae-covered rocks. However, other nudibranchs are carnivorous, and some prey upon other species of nudibranchs.

Found in tropical and temperate seas, sea hares are gastropods closely related to nudibranchs. They lack an external shell but do have internal remnants of the shells possessed by their ancestors. Sea hares graze on algae as their primary food source. They can release a noxious ink made by special glands.

Each individual sea hare, like the nudibranch, possesses both male and female reproductive organs. Sea hares often mate in large groups. They lay their eggs in masses that look similar to balls of yarn or strands of lace. Inside each strand of the mass are thousands of eggs. One sea hare may deposit several of these egg masses during a single breeding season. However, most eggs end up as food for other animals.

INVERTEBRATES

How Nudibranchs Defend Themselves

Nudibranchs face an interesting set of problems in defending themselves. Many are brightly colored and do not blend in well with their environment. Unlike many other mollusks, they do not have a protective shell, and usually they are slow crawlers and poor swimmers. But like every other animal, they must defend themselves if their species is to survive.

Many nudibranchs emit chemicals that make them taste bad. Many juvenile fish will readily bite a nudibranch, but once they get a taste of it, they spit it right out. Usually the nudibranch emerges unharmed.

In addition, some nudibranchs are capable of stealing the potent stinging cells of some corals, sea anemones, and hydroids to use in their own defense. The stinging cells are extremely sensitive, and it is almost impossible to touch a coral, sea anemone, or hydroid without causing them to fire. Somehow, though, the nudibranchs eat the stinging cells, and some of them remain unfired. The cells end up on the nudibranch's back as a discouragement to predators.

Opposite page: *A red rose on the ocean floor turns out to be the eggs of a nudibranch secreted in gelatinous folds.* This page, top: *Sea hares differ from the nudibranchs in that they have a pair of swimming fins that fold over their backs when not in use.* Above, top and bottom: *The different types of nudibranchs are separated by the location, number, and style of the ornamentation on their backs.*

INVERTEBRATES

Clams, Scallops, and Oysters

The compressed bodies of clams, scallops, and oysters reside inside two hinged shells. A tough ligament holds the two shells together. The muscle that opens and closes the shell attaches to the ligament. The hard shell can clamp tightly shut and provides good protection for the soft tissue within.

Some clams, scallops, and oysters spend their adult lives buried in sand or mud. They extend tubelike siphons into the water, using one to draw in food and oxygen and another to remove wastes. In clams, the siphons are at their posterior end.

Not all species bury themselves. Perhaps the most notable exception is the giant clam of the Indo-Pacific. These clams can reach a length of three feet or more and can weigh in excess of 400 pounds. Their mantles, like those of many clams, are beautifully colored in shades of iridescent blues, greens, and browns. Giant clams begin their lives in free-swimming larval stages, but attach to the seafloor as adults.

Other especially beautiful clams include the Caribbean antillean file shell and the giant clam's smaller cousins of the Indo-Pacific. The red mantle of the Caribbean clam has a fringe of long white tentacles. Like the giant clams, many smaller Indo-Pacific clams have spectacular mantles as well.

Rock scallops also provide a visual treat for divers and snorkelers. Close inspection of a scallop's mantle reveals a series of small bright spots. These spots are actually light-sensitive receptors that act as a simple eye. When scallops detect a sudden change in the light, they often react by closing their shell. In their larval stage, rock scallops are free-swimming members of the plankton community. As adults, however, they cement themselves to temperate water reefs with strong threads.

This page, top: *Numerous eyes and short tentacles are clearly visible along the edges of this swimming scallop from Vancouver, Canada.* Below: *A community of coccina clams at Cape Hatteras, North Carolina, extend long siphons to draw water inside their shells.* Opposite page: *The orange tentacles and fiery-red gape give this flame scallop its common name.*

INVERTEBRATES

INVERTEBRATES

Squid, Cuttlefish, and Octopi

Squid, cuttlefish, and octopi are often referred to as cephalopods, or head-foots. All species have a prominent head and attached arms or tentacles equipped with suckers. Squid and cuttlefish are similar to one another, but the fins of cuttlefish run almost the entire length of their bodies while the fins of squid usually do not. Cephalopods are among the most sophisticated and intelligent marine invertebrates.

This group contains the largest of all invertebrates, the giant squid. It can reach a known length of up to 45 feet, but specimens are rare because they live in extremely deep waters. Sperm whales feed on giant squid. The giant Pacific octopus inhabits rocky shores off California, Oregon, Washington, and Canada. They are quite shy by nature, but they are also large enough to get any diver's undivided attention. Forty to fifty pound specimens are common, and some measure 600 pounds and 16 feet. The bodies of other species are less than one inch across.

Squid, cuttlefish, and octopi do not have external shells. Squid and cuttlefish have the remnants of an internal shell while octopi have no internal shell at all. All cephalopods have sophisticated nervous systems. Their sight and sense of touch are especially well developed. Octopi even have keen chemical receptors on the ends of their arms.

Squid, cuttlefish, and octopi are excellent swimmers that have a natural jet propulsion design. They swim by taking water into an area called the mantle cavity and then rapidly forcing the water out a directable tube called a siphon. The force propels them through the water in the opposite direction in which the siphon is pointing.

Most cephalopods are masters of camouflage. They can rapidly change their color by altering special cells in their skin. In addition, most have glands that produce an inky substance they can release at will. The ink clouds the water, forming a distraction in front of predators. The ink may also block the chemical receptors of some predators like moray eels.

As a rule, cephalopods are voracious predators. They have sharp parrotlike beaks, and many species inject venom when they bite. The bite of one species, the small blue-ringed octopus of Australia, can be fatal to humans.

Above: *A pair of cuttlefish swim slowly along, arms trailing, each propelled by a fin that runs around its body like a short tutu. When necessary, these jet-propelled animals can move quickly.* Opposite page, top: *Octopi are excellent swimmers when they have to be, but they spend most of their time dwelling on the bottom.* Opposite page, bottom: *The large, glassy eyes and splayed tentacles make this squid easy to identify, even from a distance.*

INVERTEBRATES

In cephalopods the sexes are separate. Squid often spawn in large groups, and in many species the adults die soon after they mate.

Octopi lay eggs in clusters that look like a bunch of grapes. For a month or longer, the females guard and clean the eggs diligently until they hatch. The females do not feed when they are tending the eggs, and they usually die shortly after the young hatch.

The Masters of Camouflage

Capable of changing their color and shape within fractions of a second, octopi are masters of camouflage. An octopus can be dark brown and smooth one moment, cherry red and knobby the next, and only a moment later be bright blue and ruffled. In addition, octopi do not have any bones in their bodies. They also lack the hard shell found in other cephalopods. Their bodies are extremely pliable. Their flexibility allows even large specimens to crawl through very small openings.

Octopi not only can match the brightness and contrast of their surroundings, but are also able to match the specific colors of their immediate surroundings using pigment cells called chromataphores. By using muscles to change the shape of the chromatophores, octopi are able to change the color of their skin.

At times an octopus's body will literally ripple with color changes. The rippling is common during elaborate courtship rituals and when the octopus is alarmed.

INVERTEBRATES

CRUSTACEANS

Almost one million species, over 75 percent of all animals, belong to the phylum Arthropoda. Many are land-based insects, but many play major roles in the ocean environment. Marine arthropods such as lobsters, crabs, shrimps, and barnacles are all types of crustaceans.

Crustaceans make up a large portion of the zooplankton. Many of these free-swimming organisms are microscopic, but their populations are often extremely dense. Collectively they play significant roles in a great many oceanic food chains. On the other end of the spectrum, giant spider crabs found in some temperate seas have a leg span of six feet or more. Many crustaceans are not always evident to underwater explorers, but those who are patient will soon discover some type of crustacean almost everywhere they dive.

All crustaceans live inside a hard structure called an exoskeleton, a feature that provides both advantages and disadvantages. The hard exoskeleton offers valuable protection, and it gives muscles a place to attach. The primary disadvantage is that for growth to occur, the animal must go through a molting process that sheds the old shell and forms a new, larger one. Molting can take hours or nearly a full day, and the animals are left very vulnerable to predators during the process. Crustaceans molt less frequently as they get older.

All arthropods have segmented bodies, a feature that is very easy to see in the jointed legs of lobsters and crabs. As a group, arthropods show amazing diversity and specialization in their appendages. Examining a species closely often can reveal where it prefers to live and what kinds of animals it prefers to hunt. Some arthropods spend their lives as bottom dwellers. Others drift with plankton or attach to host animals as parasites. Most arthropods have well-developed senses of taste, touch, and sight.

The sexes usually are separate in arthropods. After mating, some females brood their eggs. Once the eggs hatch, however, they drift as plankton while going through a series of rapid molts. The young often go through many stages before emerging as mature animals. The juveniles of most species bear little or no resemblance to the adults. The process is similar to the metamorphosis that caterpillars undergo.

Opposite page: *Claws at the ready, a Hawaiian lobster cautiously pokes its head out from a crevice in the reef.* Above: *Sally lightfoot crabs earn their name as they scamper over rocks on an island in the Galápagos.*

INVERTEBRATES

Lobsters

Lobsters are generally shy and retiring during the day, but under the cover of darkness, they roam the bottom looking for food, mostly fish, mollusks, and other invertebrates. Lobsters often gather together in caves or crevices for shelter. A diver might see only one antenna protruding out of an opening in a rock, but a closer look reveals dozens of lobsters crammed into a tiny crevice.

Some, but not all, lobsters possess large claws. The New England lobster has a pair of large claws used in defense and in capturing and crushing prey. Some species without claws have a number of sharp spines, like the California spiny lobster and the Caribbean spiny lobster.

Slipper lobsters are less familiar to most people. Nonetheless, they are part of many coral reef, rocky reef, mud, sand, and grass communities. Slipper lobsters lack large claws and long antennae. They are short and flat with jointed legs. Some slipper lobsters are quite colorful, while others are drab.

Crabs

Some crabs such as spider crabs, blue crabs, and king crabs live completely covered by their hard exoskeletons. However, only the front portion of hermit crabs have that kind of protection. Their abdomens are soft and vulnerable to attackers. To protect themselves, hermit crabs crawl inside the shells of deceased snails. The newfound shells cover the soft abdomens and protect the hermit crabs vulnerable body parts. Small hermit crabs sometimes take the discarded shells of larger hermit crabs, and the large ones take the shells of still larger hermit crabs.

Several species of crabs cover their outer skeletons with sponges, anemones, hydroids, algae, and other living organisms as a means of camouflage and defense. These crabs are commonly called decorator crabs.

Some species of crabs become involved in long courtship rituals. The male carries or stays with the female until she molts. The male then places the soft-bodied female underneath him so that she has protection and they can successfully mate.

Opposite page, top left: *The long antennae characteristic of more familiar lobsters are modified into broad, flattened plates on the head of a slipper lobster.* Bottom left: *Divers that try to pull spiny lobsters from their lairs by their antennae are rarely rewarded with more than the broken appendage.* Bottom right: *Like all the decapod crustaceans, lobsters have ten walking legs.* This page, top: *A hermit crab, living in a gastropod shell, has decorated its home with purple anemones. When the crab grows bigger, it will need to find a larger shell and begin decorating all over again.* Above: *Sally lightfoots, like most crabs, make their own shells, which must be shed for the animal to grow.*

INVERTEBRATES

Shrimps

Worldwide, there are more than 2,000 species of true shrimps. All of them have ten walking legs, and their elongated bodies have a segmented tail and a head and thorax that are fused together. Some shrimps are excellent swimmers. For short, quick motion, they rapidly pull their tails underneath their body and propel themselves backward. It is the tail muscles that humans love to eat as shrimp cocktail.

Shrimps are very common in most marine settings, although they often go unnoticed. At night, however, pairs of eyes glow almost everywhere a diver points an underwater light. Closer inspection usually reveals that the eyes belong to shrimps.

In many reef communities, pistol shrimps and snapping shrimps are common. If divers pause quietly for a moment, they will soon hear the crackling and popping of these shrimps as they snap their claws shut on their prey.

Mantis shrimps use their razor sharp claws and quick reflexes to impale small fish and crack open hard-shelled prey such as mollusks. Their claws fold back out of obvious sight like the blade of a pocket knife, but they can be unleashed in a split second. Handling a mantis shrimp can be a serious mistake; their powerful claws can cut fingers and even sometimes smash glass jars meant to hold them in captivity. The name mantis shrimp comes from their resemblance to the insect called the praying mantis.

Many species of shrimps provide cleaning services to a variety of fish. Often the fish literally line up near the shrimps' home waiting to be cleaned of parasites, dead tissue, and bacteria. These cleaning stations are usually easy to find because of all the activity around them. Pederson's cleaning shrimps are prominent cleaners in Caribbean waters, while red rock shrimps do the job in reef communities along western North America. At times, some shrimps will even enter the mouths of predatory eels to clean and come away unharmed.

This page, above: *A small cleaner shrimp waves its long, white antennae to signal prospective fish clients.* Top: *A spotted cleaner shrimp in the Caribbean also waits for business.* Opposite page: *Other Caribbean shrimp try to avoid the same predators freely visited by the cleaners.*

INVERTEBRATES

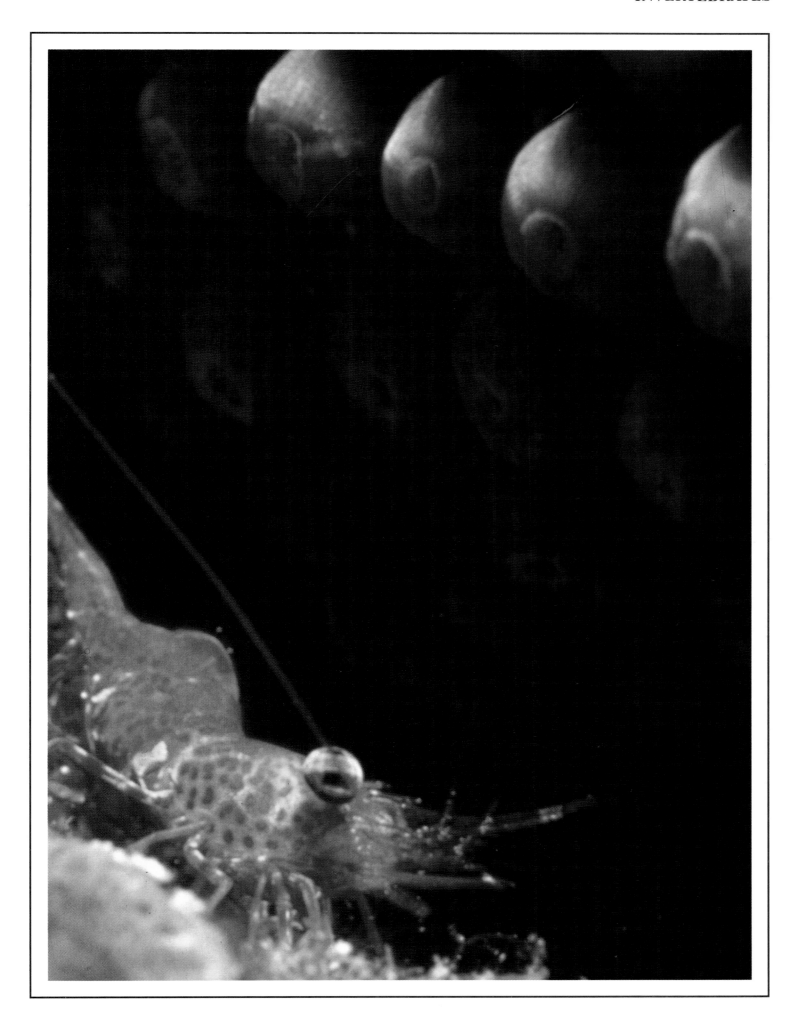

INVERTEBRATES

Barnacles

Almost everyone who owns a boat in salt water is far more familiar with barnacles than they'd like to be. Barnacles are infamous for attaching themselves to boats in large numbers. They also settle on almost any hard surface—rocky reefs, corals, pier pilings, logs, and whales, turtles, and other living animals. Special glands in the barnacle produce a strong, bonding adhesive. In their larval stage, barnacles are free swimming; as adults, they permanently attach in large numbers to anything that provides a safe surface.

Once attached, the barnacle extends its cirri, feathery appendages that reach into the water and grab drifting plankton. As it feeds, a barnacle's cirri are constantly moving, stretching out to the water and then back down to the mouth.

Barnacles need steady water movement to bring them food, so they normally colonize areas with regular currents or tides. Those species that settle on other animals or on drifting surfaces such as logs depend on their bearer's mobility to find new food supplies.

Some barnacle species that live near the seashore must cope with the tides. They settle on rock-strewn intertidal regions. These areas provide a solid surface for attachment but are regularly exposed to the open air due to the tide's flow. The barnacles have adapted to withstand regular but brief exposure to sun and air. When the tide is out, they withdraw into the protection of their shells. When the tide covers them, they once again extend their cirri and feed.

This page, top: *Only the top of goose barnacles are covered in hard plates. A fleshy stalk, or neck, that attaches the animal to a hard surface is unprotected.* Above: *The feathered fingers of an acorn barnacle are caught in the act of filtering water for food particles.* Opposite page, top: *An isopod from the Antarctic looks like a series of rough plates with legs. Small antennae, however, adorn its head.* Below: *Other isopods look like they are from another planet, but they too are made up of a series of sections, some of which have pairs of legs.*

Copepods, Isopods, and Krill

Copepods and isopods live in a wide variety of habitats. Many copepods are microscopic, living among the plankton where they feed on small plants, animals, and decaying organic matter. Isopods are usually much larger than copepods, and most are visible with the naked eye. Many copepods and isopods are parasitic, living on or in various fish and invertebrates.

Krill are tiny crustaceans, usually only one or two inches long. Despite their size, they are one of the world's most important food sources. They form a large part of many creatures' diets, from numerous invertebrates to some birds, seals, and even many of the great whales. Floating in the oceans as a major component of plankton, they sometimes form swarms of 450 square feet or more in Antarctic waters.

INVERTEBRATES

112

ECHINODERMS

All echinoderms live exclusively in the marine environment. The group includes sea stars, brittle stars, basket stars, sea urchins, sand dollars, sea cucumbers, and feather stars. The name echinoderm roughly means "spiny-skinned"; many have obvious spines or wartlike projections.

All echinoderms display five-sided, or pentamerous, symmetry. The symmetry is very easy to see in sea stars and brittle stars. Most species possess five arms that radiate from a central disk. At first glance, sea cucumbers and sea urchins seem an exception to the rule, but their bodies do show five-sided symmetry.

Almost all echinoderms are bottom dwellers. They live in all oceans, from the cold polar seas to warm tropical waters. Echinoderms range in size from sea cucumbers only one inch long to sea stars almost three feet wide.

A unique water vascular system controls the hundreds of tube feet some of the mobile species use to move. This hydraulic system also assists some species, especially those that are stationary, in capturing prey. Some echinoderms have small pincerlike organs called pedicellariae for defense, capturing food, and cleaning themselves.

In almost all echinoderms, the sexes are separate; most spawn by releasing sperm

Opposite page: *This blue sea star has a smooth texture, not unlike leather.* This page, top: *The warty surface of a sea cucumber is studded with short, dull spines.* Above: *The spines of this sea urchin are short for a sea urchin, but sharp enough to inflict a wound. This animal is releasing eggs or sperm into the water.*

Brittle Stars and Basket Stars

While brittle stars closely resemble the more familiar sea stars, it is easy to tell the two groups apart. The brittle star's arms radiate from a very obvious central disk. Sea stars' arms do not stand out so sharply from the central disk. Brittle stars' arms are also usually much thinner and more flexible than the arms of sea stars.

Some brittle stars actively hunt, while others prefer to scavenge for a living. Some are filter feeders, taking their nutrition from plankton. During the day, most species that live in shallow water bury themselves in the sand, under ledges, or in other debris, or they hide in sponges. Many divers do not even realize brittle stars exist until they make a night dive. At night, shallow-water brittle stars come out in number and begin their search for food.

Basket stars, as their name suggests, look similar to woven baskets, especially when their arms unfurl during feeding. Like brittle stars, shallow-water basket stars usually seek cover during the day, retracting their thin, branched arms. At night, they extend their arms and trap prey. Basket stars prefer a variety of plankton and small crustaceans.

Below: *Brittle stars can move quickly from crevice to crevice by writhing their five hairy arms.* Opposite page: *The five-sided symmetry of a basket star is hard to see among its many branched arms.*

INVERTEBRATES

117

INVERTEBRATES

This page, top: *The five sections of this sea urchin from the Galápagos are evident when viewed from above.* Above: *The spines on this variety of urchin offer little defense, so it camouflages itself with debris.* Opposite page: *Sea urchin spines are quite robust in some species, meant more to deter than to injure predators.*

Urchins

Any beachcomber, snorkeler, or scuba diver who has made a personal acquaintance with a sea urchin is unlikely to ever forget it. All urchins have spines. In some species, the spines are long, thin, and razor sharp. Some can easily penetrate human skin, and removing them usually proves difficult. Other urchins have dull stubby spines.

There are actually two groups of urchins. The regular urchins, commonly called sea urchins, have the medium to long spines. Irregular urchins like heart urchins, sea biscuits, and sand dollars have shorter spines and a slightly different body structure. Most sea urchins inhabit solid or rocky environments, but the irregular urchins prefer sandy sea beds where they often bury themselves.

Though the urchins' spines provide a formidable defense, fish, crabs, snail, other echinoderms, and sea otters still prey on them. The California sheephead, for example, can pry urchins from the rocks and crush them. Most urchins feed on algae. Many are scavengers, eating almost anything they come across.

The sea urchin's skeleton, or test, catches the eye of many a collector. The unusual skeletons consist of 20 rows of thin plates with small holes. Sand dollar tests also end up in the personal effects of many beach goers. Sand dollars are circular and very flattened.

INVERTEBRATES

119

INVERTEBRATES

120

Sea Cucumbers

By appearance, sea cucumbers don't seem to fit in with other echinoderms. They have sausage-shaped bodies, and even careful inspection can leave any hint of five-sided symmetry undetected. If you dissect a sea cucumber and look lengthwise down its body, though, you will clearly see its five symmetrical sections.

While the general body shape of most species is similar, their coloration varies. Some are drab brown to dark green, while others are snow white to bright orange. Wartlike projections or spines often cover their bodies.

Most sea cucumbers are either deposit feeders or suspension feeders. The deposit feeders use their adhesive tentacles to pick up decaying organic matter. The suspension feeders trap small particles on their outstretched tentacles.

When relaxed and feeding, a sea cucumber is soft and elongated. When it feels threatened, it can shorten its body. If the threat intensifies, some resort to bizarre defense mechanisms. Some species expel a sticky mass of tubules that sometimes contains toxins to repel predators. Many sea cucumbers can actually expel some of their own internal organs as a way of escaping predators. The strategy works because of the sea cucumbers' amazing regenerative ability. They can quickly replace the lost organs.

Opposite page: *Five rows of tube feet delineate the five equal sections of this blue and red sea cucumber. Branched prehensile tentacles around the mouth search out and pick up food items.*
Right: *Distraught sea cucumbers can eject a mass of tubules to ward off their agitators.*

Regenerating Lost Body Parts

Many echinoderms possess an amazing ability to regenerate, or regrow lost body parts. For example, if a sea star loses an arm to a predator, it will quickly grow a new one. Even more astonishing, if just a tiny portion of the central disk remains with the lost arm, the arm will develop into a completely healthy five-armed sea star. Brittle stars can also regenerate this way.

Regeneration has even become a key feature in many species' defense strategies. Many brittle stars and sea stars have arms that pull off quite easily. In many instances, giving up an arm is less dangerous than fighting to save it.

Sea cucumbers can even regenerate their digestive and respiratory systems. When attacked, many sea cucumbers expel some of their organs to repel predators.

Crinoids

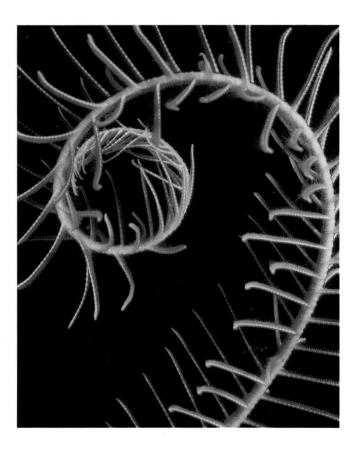

This page, top: *A feather star's arm is delicately coiled when not feeding.* Above: *Some feather stars can swim through the water or walk along the bottom, but this one is attached to the bottom by a stalk.* Opposite page: *Usually at night feather stars extend all their arms into the water to filter out small particles.*

Today, the ocean houses about 600 species of crinoids. The fossil record shows that several thousand species of crinoids existed over 300 million years ago. Crinoids are the most primitive form of echinoderms. Modern crinoids lack both spines and pincerlike pedicellariae. They do, however, have a series of plates that cover their bodies.

Some people call crinoids feather stars or sea lilies. Many species sport brilliant hues of burgundy, cherry red, bright yellow, green, and an almost endless variety of mottled color combinations. Still others are relatively drab, and these specimens often live side by side with their vividly colored brethren.

As adults most feather stars move about using their arms and a ring of appendages called cirri. In many species, the five arms quickly fork several times, forming branches. The branches support numerous thin tube feet that collect plankton and other food that drifts past.

Sea lilies attach to the bottom by a stalk. To capture plankton, they unfurl their arms and sway back and forth. Observant divers can watch exposed crinoids rhythmically move their arms back and forth through the water as they capture their food.

INVERTEBRATES

VERTEBRATES

There are approximately 46,000 species of vertebrates living today. They all possess a skeletal system including a backbone or spinal column, well-developed organ systems, and well-developed senses. This familiar group includes fish, reptiles, mammals, birds, and amphibians.

In the marine world, vertebrates are among the most obvious and well-known creatures. They tend to be larger and more active than most other ocean dwellers, and often they dominate their habitats.

As a group, marine vertebrates are quite diverse; they include snakes and turtles, sharks and countless fish species, and whales, dolphins, and seals. They thrive in literally every ocean environment, and their behavior and development tell some of the most remarkable stories the ocean has to offer.

Opposite page: *An imperial angelfish passes over a reef in search of food.* This page, top: *A sea snake glides across the ocean floor.* Above: *A sea horse clings to the blade of an underwater plant with its prehensile tail. What do these animals have in common? They all have backbones.*

VERTEBRATES

TUNICATES

Tunicates, or sea squirts, form a distinct and unusual group of marine animals. Worldwide there are over 1,300 species, and they occur in all seas. Tunicates serve as an evolutionary link between invertebrates and vertebrates.

This page, top: *Gelatinous sea squirts, also called tunicates, look like clumps of colored grapes.* Above: *When feeding, each individual tunicate filters water through an incurrent and an excurrent pore.* Opposite page: *Though simple in appearance, these colonial animals are rather high-ranking marine creatures, taxonomically speaking.*

Tunicate larvae have a completely different appearance than the adults. The larva resembles a small tadpole. It has an enlarged head that quickly tapers to a thin tail. Most importantly though, a notochord is obviously present in the tail. This flexible mass of cells runs the length of the body and gives the animal structural support. Notochords also appear in all vertebrates, usually early in development before the spinal cord forms.

Sea squirt larvae soon settle to the seafloor where they attach by using adhesive pads located on their heads. The tail and notochord disappear, the body becomes tube or barrel shaped, and two external openings appear. The entire animal becomes encased in a jellylike or leathery case or tunic—hence the name tunicate. Consuming microscopic organisms, adult tunicates are extremely efficient filter feeders.

As adults, some tunicates are solitary. Many solitary adults have an indistinct, globular shape. Their size varies, but most are less than six inches tall. Some species bear vivid shades of purple, yellow, and iridescent blue. Others are a rather drab off-white or brown. Still other species live a colonial existence. Large colonies often cover several square feet.

Free-living tunicates, known as salps, spend their adult lives drifting in the open sea. Some varieties of salp live in colonies called salp chains. Often curled into attractive spiral patterns, salp chains look somewhat like interconnected, translucent Christmas tree ornaments. Most are only a few feet long, but they can measure as long as 40 feet.

VERTEBRATES

VERTEBRATES

128

MARINE REPTILES

Less than 70 of the world's approximately 7,000 species of reptiles are marine inhabitants. The saltwater species include one species of lizard, seven species of sea turtles, and roughly 55 species of sea snakes. Aquatic reptiles probably evolved from their terrestrial-based cousins. They have lost many adaptations found in their land-dwelling relatives but have developed other features that enable them to survive the sea.

All reptiles are cold-blooded, which means most of their body heat must come from outside their bodies. Water conducts heat far more quickly and efficiently than air does, and most marine reptiles require the warmth of tropical or temperate seas.

Marine Iguanas

Found only in the Galápagos Islands off Ecuador, the marine iguana is the world's only sea-going lizard. Marine iguanas are fierce-looking creatures, but they have a shy and inoffensive demeanor. These lizards spend most of their lives on rocky shores, but nearly every day they enter the sea to feed.

Even though the Galápagos Islands are very near the equator, the water temperature is surprisingly cold due to the Peru Current. This current originates near Antarctica and flows northward along the Pacific coast of South America. Near Ecuador it deflects westward toward the Galápagos Islands. The current cools the island waters, making it far colder than most tropical oceans.

Opposite page: *Sea turtles were once numerous in near-shore waters, but their numbers have declined due to heavy fishing. Even turtle eggs are collected in some areas.* This page, top: *A rare sight anywhere else in the world, this marine iguana perches defiantly atop a rock in the Galápagos Islands.* Above: *Iguanas get chilled in the cold Pacific water and must come ashore to warm in the sun.*

VERTEBRATES

This page, top: *Two male marine iguanas, recognizable by the reddish hue of their skin during the mating season, may battle to determine which will dominate.* Above: *Wearing leftovers like a wig, a large iguana emerges from the sea after feasting on a diet of green algae.* Opposite page, top: *A sea snake devours its primary food, a small fish.* Bottom: *Rather than hiding, many sea snakes advertise their presence with bright colors so predators know to stay away from these highly venomous creatures.*

These water temperatures present a problem for the cold-blooded marine iguanas. The iguanas consume algae found in the cold, off-shore waters. To survive the cold, they must maintain a high body temperature. They solve their problem with a behavioral adaptation. In the early morning, these reptiles assume a flat basking posture on the coastal rocks and catch the tropical sun's warming rays. By midday, the sun has sufficiently warmed their bodies, and they can enter the water to feed.

Marine iguanas are excellent swimmers and divers. Females and juveniles tend to feed close to shore, but the larger males easily penetrate large surf and dive 30 feet or more to graze. To reduce their need for air and increase their dive times, marine iguanas significantly slow their heartbeat when they submerge. In addition, their metabolic rate is comparatively slow. These adaptations are common in marine reptiles.

When the iguanas finish feeding, they return to land. There, the reptiles bask in the afternoon sun, raising their body temperature so they can survive the cool nights. The following morning, they again warm themselves so they can venture into the sea.

Iguanas' skin is normally a drab black or brown with a few gray, green, or reddish splotches. During the mating season, which varies from island to island, the males become mottled with bright green, red, or orange spots. Males do not control females, but instead dominate other males within a given territory. During courtship, the male elaborately circles the female while nodding his head. After a successful courtship, the females find sandy areas where they lay two to four eggs that will hatch in two to four months. Hatchlings fall prey to a variety of birds including the Galápagos hawk, which also hunts the adults.

Sea Snakes

Sea snakes are marine reptiles that have an entirely aquatic lifestyle. There are about 55 species, all native to shallow Indo-Pacific waters where they feed on fish. Many species such as the yellow-bellied sea snake display attractive color patterns, although others such as the olive sea snake are drab green or light brown.

Sea snake venom is considerably more potent than that of their terrestrial cousins, and many people fear them. Certainly these reptiles deserve respect, but they pose no great threat to swimmers and divers who don't try to handle them. Sea snakes show little unprovoked aggression toward humans, with the possible exception that some species may attack during their mating seasons. Most bites of humans occur when people attempt to free sea snakes that have become entangled in fishing nets.

The chemical composition of sea snake venoms varies from species to species. All, however, are fast-acting neurotoxins that quickly incapacitate prey. Sea snakes are air breathers, as are all reptiles, and their poisons prevent long, energy-consuming battles underwater.

Sea snakes are excellent swimmers. Some can remain underwater for several hours, and dives often last 15 minutes or more.

VERTEBRATES

132

Sea Turtles

Turtles first appeared on earth over 200 million years ago. Only about 50 million years later, not long on the evolutionary scale, the earliest known sea-going turtles appeared. The largest living species is the giant leatherback turtle at about six feet and 1,000 pounds. The two smallest species, the Kemp's ridley and olive ridley turtle, are just over two feet long and weigh about 80 pounds. The remaining four species are the green turtle, hawksbill, loggerhead, and Australian flatback green turtle.

Sea turtles populate tropical and temperate seas around the world. They range as far north as Newfoundland and as far south as the southern tip of Africa. The range of each species varies considerably. The Australian flatback resides only in Australian tropical waters, and the Kemp's ridley inhabits the waters of the Gulf of Mexico. By comparison, loggerheads live in temperate and tropical seas around the world. Leatherbacks range as far north as Labrador and Alaska, farther north than any other turtle.

The adults of most species inhabit shallow coastal waters, bays, lagoons, and estuaries but sometimes venture into the open ocean. The exception is the leatherback, which stays in the open ocean when not nesting. Some turtles migrate great distances, but surprisingly, migration habits can differ among different populations of the same species. For example, some green turtles nest and feed in nearby areas, but other populations migrate close to 1,500 miles between their nesting beaches and feeding areas. Leatherbacks migrate farther than any other species, often covering a distance of 2,000 miles. By contrast, hawksbills rarely migrate.

Sea turtles differ somewhat from their land-based cousins. Their limbs and head cannot retract into their shells, and their flippers have adapted to assist in swimming. Their paddlelike front flippers provide thrust, while their back flippers function like a rudder. Sea turtles are excellent swimmers, and though they usually cruise quite slowly, they may be capable of short bursts exceeding 20 miles per hour.

Opposite page: *Hundreds of loggerhead turtle hatchlings, each only a few inches long, make a mad dash for the surf on a beach in South Carolina. Few will survive to adulthood.* Above: *Although designed for chewing plants, the finely serrated jaws of an adult green sea turtle can deliver a nasty bite.*

Leatherbacks can dive as deep as 3,300 feet. Typical of cold-blooded creatures, sea turtles have slow metabolic rates, a feature that allows them to dive up to 45 minutes. When resting as opposed to actively swimming, green turtles can stay submerged for five hours. Some green turtles can hibernate underwater for months.

The diet of sea turtles varies from species to species and covers the full spectrum. Some species are herbivorous, others carnivorous, and still others eat both plants and animals. Green sea turtles change their diet as they age. During their first year or so, they hunt small mollusks, crustaceans, echinoderms, and some fish. After that, they feed primarily on marine plants.

Sea turtles mate in the ocean but lay their eggs on shore. The females crawl ashore, usually during the night at high tide, and dig a nest where they can deposit their eggs. They lay their eggs at high tide to be sure the nest is above the waterline.

Interestingly, the offspring's sex remains undetermined at the time of fertilization. Instead, the nest temperature determines the sex. Lower temperatures, which usually result from deeper nests, produce males, while higher temperatures yield females. A nest's eggs will normally hatch at different times. If temperature conditions change, the same nest can produce both males and females.

Incubation time varies from 40 to 70 days. The young usually hatch at night and quickly scramble toward the sea. Life is unkind to the hatchlings; as many as 90 percent do not survive the first year, succumbing to predatory birds, fish, crabs, and killer whales. Those that do survive live to be at least 20, and perhaps 50 years old.

This page: A female loggerhead sea turtle returns to the beach where she hatched to dig a hole in the sand and deposit over a hundred leathery eggs. Opposite page: Sea turtles spend almost their entire lives in the water.

VERTEBRATES

VERTEBRATES

BONY FISH

Scientists have currently documented more than 20,000 species of fish, and some maintain that at least 8,000 species remain unclassified. In sharp contrast to reptiles and mammals, more fish species are living than are extinct. By far, most species live in the world's oceans and estuaries. The remainder inhabit fresh water. Of the marine species, the greatest variety occurs in tropical seas. The fewest species exist in polar seas.

The Design of Bony Fish

There are two classes of fish with jaws, the bony fish and the cartilaginous fish. As the names imply, bony fish have a skeleton made of bone, and cartilaginous fish — the sharks, rays, skates, and chimaeras — have a skeleton made of cartilage.

Some basic physiological features are common to all fish. They all have gills throughout their lives, and most have fins. Many have scales, though their size and shape vary greatly from species to species. All fish secrete mucus from their skin that blocks bacteria and fungus and reduces water friction.

Many fish have spectacular colors. The colors assist in camouflage, advertise a fish's presence, or provide some form of communication to other creatures. Specialized pigment cells called chromatophores produce most of these colors. The cells have a direct link to a fish's nervous system, allowing the fish to contract or expand the cells and change the color of each one. Color changes can be brought about in a matter

Opposite page: *Many tropical reef fish, such as this juvenile king angelfish, sport ornate markings in neon colors.* This page: *These two lizardfish demonstrate the species' remarkable knack for camouflage. They can alter their color to blend in perfectly with almost any background.*

VERTEBRATES

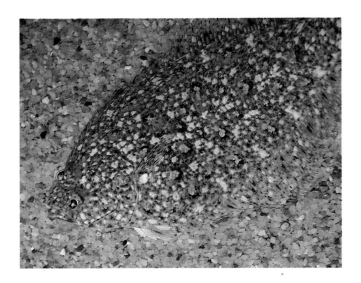

of seconds. Many fish alter their color to match their surroundings. Lizardfish, flatfish, and trumpetfish frequently change their coloration; many other species rarely, if ever, alter their color.

Left: A well-camouflaged sanddab from Monterey Bay, California, spends most of its life lying on its side in the sand. Opposite page: A ghost pipefish from New Guinea has a head like a pipefish or sea horse, but its body is more robust.

Bony Fish vs. Cartilaginous Fish

Science has established two major divisions of fish—those with jaws and those without. There are two types of jawed fish, and they make up the vast majority of fish species. They are cartilaginous fish or Chondrichthyes, and bony fish or Osteichthyes. Cartilaginous fish include sharks, rays, and skates, fish with skeletons made of cartilage, not bone. Most fish, in fact more than 18,000 of the world's 20,000 species, have skeletons made of bone.

While the bony and cartilaginous fish are classified by their bone structure, the two groups differ in other ways too.

1) Bony fish have bony scales that cover their skin, while cartilaginous fish have a different kind of scale. The skin of sharks, skates, and rays is actually a sheet of small toothlike structures called denticles. Under a microscope, the denticles look much like a series of miniature teeth.

2) Most bony fish possess gills on either side of the head that have a bonelike covering or operculum. The operculum pumps oxygen-filled water over their gills. Cartilaginous fish possess five to seven gill slits on either side of the head, but they lack an operculum.

3) A swim bladder is typically present in bony fish but lacking in cartilaginous fish. This internal organ retains or releases gases from the bloodstream. Because the gas is lighter than water, more gas makes the fish more buoyant. By precisely controlling the amount of gas in the swim bladder, bony fish can float, hover, or sink.

4) Bony fish typically reproduce by spawning; either both sexes release sperm and eggs into the water simultaneously, or the male releases sperm over a nest of eggs. Sharks, skates, and rays reproduce through copulation or internal fertilization.

5) In terms of their overall anatomy, the mouths of bony fish are usually at the very front of the body, and they have symmetrically shaped tails. The mouths of cartilaginous fish generally rest just below the head, and their tails are often asymmetrical in design.

VERTEBRATES

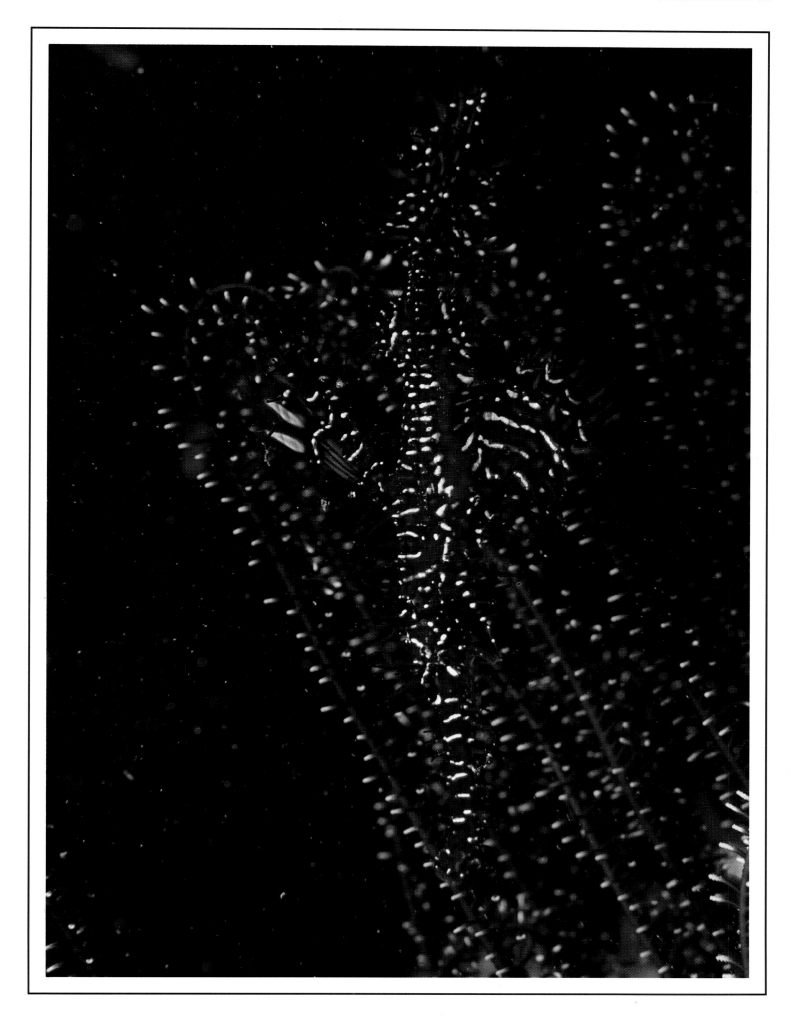

VERTEBRATES

Fish Senses

Fish have well-developed eyes that are very similar to the eyes of humans. In many species, each eye moves independently and rotates over a great range for a wide field of view. In a world where predators strike silently from any direction, this can mean the difference between life and death. Fish that live in shallow reef communities probably have good color vision, but deep-water species have few color receptors. At great depths there is little light, and the receptors have almost no value.

Water transmits sound much better than air does, and the underwater world is far noisier than many people suspect. Sounds come from a variety of sources: snapping shrimps, grunting squirrelfish and groupers, the crunching and grinding of fish that graze on reefs, and the high-pitched singing of dolphins and whales. Fish have well-developed inner ears located near the skull, but they have no outer or middle ear. People experienced in fishing know that loud noises will scare many fish, and divers are well aware that many fish react to the noise made by the bubbles of scuba systems.

For many humans, it is difficult to imagine smelling anything in water. But fish depend heavily on their well-developed sense of smell. Olfactory receptors line numerous openings along the snout. Fish such as moray eels rely on their sense of smell to help locate prey when they hunt. And some species emit chemicals with distinct odors to communicate with other members of their species.

Surprisingly, the sense of taste is not well developed in most fish. Certainly it is not absent, but studies show that taste is not as important as other senses in most fish. On the other hand, most fish are extremely sensitive to touch and will quickly flee from any contact with foreign bodies.

Do Fish Sleep?

As is so often the case in nature, there is not a definitive answer to the question of whether or not fish sleep. Some do and some don't. Some are deep sleepers, and others vary their sleep patterns from day to day and as environmental conditions change.

The common consensus holds that most open water fish such as blue sharks, whale sharks, manta rays, mackerel, and herring do not sleep. Many open water species are most active at night. While they do become far less active during the day, they do not sleep.

Most reef fish are diurnal, meaning they have an active and an inactive period every 24 hours. Parrotfish, angelfish, damsels, and butterfly fish seek cover at night so they can sleep without risking predation. Not all species retire or awaken at the same time, but members of the same species usually rest or arise within a few minutes of one another every day.

Opposite page, top: *Squirrel fish have large eyes to help them see at night when they are most active.* Bottom: *A clown triggerfish cruising over a bright Australian reef during the day doesn't need large eyes.* This page, top: *A long-nosed butterfly fish uses its snout to reach down into crevices inaccessible to other creatures as it searches for small crustaceans.* Above: *A moorish idol, with its pointed snout, also picks up tidbits of food that are out of reach for many other fish.*

VERTEBRATES

This page, top: *With its real eye hidden in a black band and a false eyespot on the trailing edge of a fin, this butterfly fish often misleads predators about the direction in which it will be heading.* Above: *Shape, color, and posture help a trumpet fish hide among the branches of a soft coral colony.* Opposite page: *Lionfish in the Red Sea stand out against the reef, but predators know to stay away from the long frilly fins trimmed in white that signal poison.*

How Fish Use Colors

The dazzling array of colors present in the world's fish seems endless. Fish vary in color from extremely drab to handsome to almost gaudy. Coloration serves a variety of purposes, and that offers probably the best explanation of why there are so many variations in color patterns.

In many cases, color enables fish to hide by helping them blend into their immediate surroundings. Certainly that is the case with camouflage artists such as halibut, turbot, and sole that change color to match their surroundings from moment to moment. Other species do not alter their color, but instead live where their coloration constantly matches their surroundings. For example, giant kelpfish reside in kelp forests where they blend in quite easily. Even the bright color patterns of many dazzling reef fish actually make it difficult to see the entire individual at once, which helps disguise the creature's movement.

Some fish use their bright coloration as a warning. Distinctively colored lionfish use their toxic dorsal spines to ward off predators. Their unmistakable coloration lets other sea dwellers recognize them as formidable enemies, making it more likely that they will leave the lionfish alone. Still other species use variations in color to distinguish males from females.

VERTEBRATES

143

VERTEBRATES

Diversity in Fish

Fish are an incredibly diverse group of animals. They vary considerably in size, shape, where they live, how they live, what they eat, and what eats them. Some fish are exceptionally colorful, others quite drab. Most, but not all, have fins. Tuna, wahoo, and jacks are true speedsters; moray eels are graceful and snakelike in their movement; other species are poor swimmers.

Fish vary in size from the Marshall Island goby, which is only 12 millimeters long when full grown, to the 40-foot whale shark, the largest living cold-blooded animal on earth. Species such as puffer fish have inflatable bodies, while the inflexible bodies of sea dragons and sea horses rest in an armored coat of rings and plates. Pipefish, trumpetfish, and needlefish have long thin bodies. Ocean sunfish are large oval disks, somewhat like overgrown frisbees with fins.

Feeding strategies vary considerably from species to species, which allows many species to coexist in the same communities. If two species pursue the same food at the same time of day in the same manner, one species will eventually outdo the other, and the loser will perish. One way that species can share a common food source is to have a different daily cycle. For example, many wrasses feed on tiny crustaceans and worms by day. When the wrasses retire for the night, squirrelfish leave their daytime shelters and hunt the same crustaceans and worms.

Bottom-feeding fish use many strategies to capture their food. Trunkfish often blow

continued on page 148

Opposite page: *An inflated puffer fish covered with spines looks like an unappealing meal to most potential predators. When not inflated, this fish is half the size and its spines lay flat.* Below: *Sea horses, wearing their scales like armor, may be the most improbable of fish. When not maneuvering with miniature fins, they anchor themselves down with their curled tail.*

VERTEBRATES

146

Why Some Fish School

Of the 20,000 fish species known to science, about 80 percent school as juveniles and 20 percent school as adults. Questions remain about why fish school despite a great deal of study. Certainly schooling behavior entails some cost. A school of fish must find a lot more food so that each individual gets enough to eat. But there are also some benefits, and they provide the best explanation of schooling's purpose.

1) Schooling can confuse predators so that it's difficult for them to pick out one target. In order for a predator to be successful, it must focus its efforts on a single creature. The flurry of activity in a school makes it difficult for a predator to target a single prey item.

2) Some predators such as tuna and jacks school to counter the schooling tactics of prey fish. Working together, the predators break the school of prey into smaller unorganized groups so it's easier for them to select a single victim.

3) The idea of safety in numbers applies to schooling fish. If a predator attacks a school, the odds for survival for any single fish are greater when the school is larger. In other words, each individual hides within the school, hoping that it will not be the one that gets eaten.

4) Schooling offers a survival strategy to many species by making it easier for them to reproduce. Forming schools makes it more likely that males and females will find each other and mate. This leads to more offspring and a greater chance that the species will endure.

5) Schooling makes swimming easier because it reduces water resistance. Each fish follows the wake of the fish in front of it, so swimming requires less energy for most of the school.

Opposite page: *Thousands of finger-sized silverlings school to create a single, large moving mass that may overwhelm predators.* Below: *Large, fast-swimming white jacks school, albeit in smaller numbers, as they search out their own prey.*

VERTEBRATES

This page, top: *Goatfish use their barbels as feelers to root in the sand for food.* Above: *A trumpetfish patiently waits for a smaller fish to wander by. With amazing speed, it opens its tubelike mouth at the tip of its snout, sucking in unsuspecting prey.* Opposite page: *A look into the cavernous maw of a Nassau grouper reveals serrated gillrakers.*

continued from page 145

jets of water into the sand to uncover small crustaceans and mollusks. Goatfish use the whiskerlike barbels under their chins to feel for their prey. Some fish steal their food, grabbing the prey uncovered by species that sift the bottom.

Filter-feeding anchovies and herring have mouths that strain food from the water. Trumpetfish and pipefish have elongated mouths that suck in their prey. And of course, there are the predators—the barracuda, seabasses, jacks, and tuna—that have many sharp teeth that help them snag their prey.

Water temperature is a critical factor in determining where fish live. Most species remain in either tropical, temperate, or polar seas; very few can survive in such vastly different water temperatures. And each species in a particular water temperature finds its own preferred habitat within that zone. Some fish live their lives in close association with the seafloor, but others spend their entire lives in the open sea.

Fish species also vary in their territorial instincts. Many species simply wander through the sea. Although they generally prefer certain areas at given times of the year, they do not defend a territory in which they feed or breed. Some of these species, like jacks, tuna, mackerel, anchovies, and herring, live in large schools, while marlin and sailfish are non-schooling species that roam the open ocean.

Like their oceanic relatives, the schooling fish in reef communities do not claim territories. However, solitary reef-dwelling

VERTEBRATES

VERTEBRATES

species such as damsels do claim territories that they vigorously defend. Damsels actually cultivate patches of algae within their territories. Some reef species are territorial throughout the year, while others are territorial only during their mating season. Surgeonfish are territorial when they are sexually immature, but not as adults.

Defending a territory by chasing away an intruder can be a very costly energy expense, and it involves some risk. Many fish have developed threat displays that help avoid confrontations. During displays, the invader and defender threaten and posture at a distance. Often, the invader vacates the defender's turf without a physical conflict.

Bony fish also show diversity in their breeding strategies. Courtship patterns and breeding rituals are elaborate in many species. Many schooling fish are broadcast spawners that mate in large gatherings. Some gather only to reproduce. Most species usually produce tremendous quanti-

Opposite page: *A school of wrasses twist and dart over the reef in an ancient mating ritual.* Below: *Two male wrasses engage in a symbolic territorial display. The confrontation will most likely end peacefully, which benefits both parties.*

VERTEBRATES

VERTEBRATES

ties of fertilized eggs to ensure that some will survive the rigors of ocean life.

Many flying fish attach their eggs to clumps of seaweed, where the eggs remain until they hatch. Garibaldis and sergeant majors lay their eggs in sheltered areas that the males defend fiercely. Jawfish go to another extreme—the males protect the eggs, holding them in their mouths until they hatch. Male sea horses and sea dragons carry the unhatched eggs on their abdomens.

Once hatched, many fish experience a larval stage during which they do not look or behave at all like the adults. For example, larval tarpon are semitransparent and drift in the open sea, while the adults live in shallow coastal communities, estuaries, and bays. Many juvenile rockfish and triggerfish drift in the open sea before settling down in reef communities. Perhaps the most fascinating transformation occurs in flatfish—halibut, sole, turbot, flounder, and sanddabs. In their larval stage, these open sea residents have one eye on either side of their heads as most fish do. As they mature, one eye migrates over so that both eyes are on the same side, and the adults settle into shallow sand communities.

Opposite page and this page: *A male sergeant major, in charge of guarding its purple eggs as they lie on the seafloor, will attack or remove any intruders in his nesting territory. It may take him several attempts to move this starfish, but his efforts will help ensure the survival of the next generation.*

VERTEBRATES

A solitary filefish (this page, top), *an elegantly patterned queen angelfish* (above), *and the countless small baitfish in this school* (opposite page) *represent what is perhaps the most diverse group of marine animals, the fish.*

The Ultimate Mid-Life Crisis

Imagine being given a biological message that commands you to change your sex. That's right, to change from male to female or vice versa. As strange as it sounds, a mid-life sex change is a common adaptation in many fish species. Wrasses commonly undergo such a change. The fundamental reason that wrasses, or any other fish, change sex is that it gives the species a better chance to survive.

In most species of wrasses, the newborn fish are either all females or both females and non-sex-changing males. All females have the capacity to become males. Terminal males are fish that were once females but became males. As a rule, terminal males are very brightly colored. Upon reaching sexual maturity, non-sex-changing males spawn in large groups, while terminal males are much more territorial and spawn with the females that enter their mating sites.

In many species of wrasses, each population has a dominant male and a dominant female. The males are extremely aggressive and control the females. If a dominant male dies or leaves the population, the dominant female changes sex. Within a matter of hours the female turned male displays the characteristic male aggressiveness toward females and patrols his new territory. Within a few days, the new male produces sperm.

VERTEBRATES

SHARKS AND THEIR KIN

Without question sharks are among the most feared creatures on earth. Few images convey the terror humans feel when we see a shark swimming toward its prey with mouth agape and razor-sharp teeth fully exposed. Many people see sharks as mindless eating machines, voracious hunters and indiscriminate feeders that become frenzied when they sense a single drop of blood. Scientists and divers who have worked with sharks know this image is incorrect. In reality, sharks are as misunderstood as they are feared.

Perhaps the first thing to learn about sharks is that there is no one animal called the shark. Worldwide there are over 350 species. While the species share many common characteristics, they are also considerably different in many ways. Sharks differ in their size, where they live, what they prefer to eat, and how they fit into nature's overall plan. A look at size shows the great variety among species. The whale shark is the world's largest fish at 40 feet and 30,000 to 40,000 pounds. In contrast, the dwarf or pygmy shark never exceeds nine inches. Surprising to many, more than 80 percent of sharks are less than 6 feet long when fully grown.

Most members of the shark family have been designed to kill, and they have been designed quite well. Their deadly efficiency can be frightening, but they are still graceful, elegant, and beautiful. Opposite page: *Marbled rays often hover over the seafloor as they hunt for crustaceans.* Above, left: *Whitetipped reef sharks may be the most common shark in the South Pacific.* Above: *A scalloped hammerhead shark is easy to identify by its oddly shaped head.*

SHARKS AND THEIR KIN

This page, top: *The eyes and nostrils of a scalloped hammerhead shark are set far apart, out on the ends of its wide head.* Above: *Although sharks are opportunistic feeders, they eat mostly at night.* Opposite page: *A blacktip shark cruises in the Bahamas.*

Sharks also differ considerably in their preferred diet. Horn sharks, angel sharks, and whitetip reef sharks feed on crustaceans, mollusks, echinoderms, and small fish, while mako sharks feed almost exclusively on squid and open ocean fish such as mackerel and tuna. Blue sharks also feed on squid and a variety of small schooling fish such as mackerel, herring, and anchovies. Adult great white sharks prefer to feed on marine mammals such as sea lions and seals. And perhaps surprising to many laymen, the two largest species, whale sharks and basking sharks, feed primarily by swallowing enormous quantities of microscopic plants and animals called plankton.

THE BIOLOGY OF SHARKS

Sharks are a very successful life form, having adapted extremely well to the demands of the marine environment. Well-developed senses and many other adaptations have made them one of the dominant predators in many marine habitats for millions of years.

Sharks and their close relatives, the rays and skates, make up a scientific grouping of fish called elasmobranchs. All elasmobranchs have skeletons of cartilage; most fish have skeletons of bone. As a group, sharks, rays, and skates have lived on earth for over 450 million years. Their fossil record dates back more than twice as long as that of the dinosaurs. Because cartilage precedes bone in the skeletal development of animals, many scientists thought that bony fish evolved from sharks. This belief contributed to the misconception that

SHARKS AND THEIR KIN

SHARKS AND THEIR KIN

sharks are primitive, simple animals that lack intelligence. Recent discoveries, though, have supported modern theories that the two groups evolved independent of each other.

The typical shark has a slender, graceful body that is slightly thicker in the middle and tapers at both ends. The snout, or nose, ranges from pointed to round to squared off. Typical sharks have a pair of dorsal fins located along the top of their back and a sickle-shaped tail whose upper lobe is larger than the lower lobe. The mouth of most species sits on the underside of the head. This general description applies to a wide range of species such as gray reef sharks, blue sharks, whitetips, sand tigers, bull sharks, lemon sharks, tiger sharks, and many more.

However, there are significant variations in body design in other species. The bodies of angel sharks and wobbegongs are greatly flattened. Mako sharks, porbeagles, salmon sharks, and great whites have nearly symmetrical tails, with the upper and lower lobes almost the identical shape and size. And the mouths of whale sharks are at the front of the head rather than underneath.

Sharks lack the internal organ called a swim bladder found in most bony fish that assists in buoyancy control. Still they are very efficient swimmers. Sharks gain their forward thrust with the back and forth movement of their long, powerful tails. They use the pectoral fins along their sides to control turns. The pectoral fins also provide lift in a manner similar to the wings of airplanes. Other fins assist in braking, tracking, and stability. Most sharks are excellent cruisers, but they tend to tire quickly if they swim at maximum speeds.

The cartilaginous skeleton of sharks also aids in swimming. Cartilage is much more flexible and lighter than bone. The increased flexibility contributes to the sharks' graceful, sinuous swimming motion. The lighter weight of cartilage helps sharks conserve energy when they swim. Because they lack a swim bladder, sharks must swim constantly to support themselves in water, and the lighter skeleton makes it considerably easier for them to do this.

Opposite page: *Bottom-dwelling angel sharks are easy for divers to approach. They are normally docile, but divers still need to exercise caution. They will attack if provoked.* Above: *The large mouth on the front of the whale shark's head is well developed for filter feeding.*

SHARKS AND THEIR KIN

Sharks have enormous livers, which also compensates somewhat for the lack of a swim bladder. Shark livers are rich in oil. Because the oil is less dense than water, it makes the shark more buoyant. In many species, the liver alone comprises more than 20 percent of the overall body weight.

Many sharks swim throughout their lives, and the act of swimming is an integral part of their respiration. Active species such as blue sharks, oceanic whitetips, and mako sharks depend on their forward momentum to pass sufficient amounts of oxygen over their gills. Most bottom-dwelling species are less active swimmers. Angel sharks, swell sharks, horn sharks, and nurse sharks can breathe without swimming. They sometimes rest on the bottom for extended periods of time.

Caribbean reef sharks often rest in caves off the coast of Mexico's Yucatán Peninsula where they are said to be "sleeping." Though the sharks are not in a true state of sleep, many of their body functions are dramatically slower than normal. And certainly they are not as aware as when they are swimming. Unusual water characteristics play a role in this behavior. The water in the caves is rich in oxygen, and when these sharks rest, they tend to face into a slight current to help oxygen pass over their gills.

Below: *The mottled wobbegong shark, or carpet shark, is well camouflaged against a sandy bottom. This shark has sensory barbels along its mouth that help it detect prey and may also act as a lure.* Opposite page: *A swell shark in 1,000 feet of water off New Guinea rests on a coral. If disturbed, these sharks can inflate with water like a puffer fish.*

The life expectancy of many sharks is surprisingly long. Studies have shown lemon sharks and piked dogfish sharks can live at least 70 years, and scientists speculate that they live 100 to 125 years in natural settings. Sharks tend to grow rather slowly, especially those species with long life spans. The piked dogfish, for example, grows at a rate of only one to two inches per year. Open ocean species such as blue sharks and mako sharks grow ten to 15 inches per year when food is abundant.

SHARK SENSES

Sharks use a combination of senses to acquire their food and to analyze their surroundings. These senses include smell, feeling, hearing, sight, taste, and the special ability to detect electrical and magnetic fields.

Most of a shark's olfactory receptors are small sacs in its nostrils. Much is made of the fact that sharks have a very keen sense of smell. Laboratory experiments show that some sharks can detect one part of blood in one million parts of water. However, just because a shark can detect the chemicals in blood does not mean it can locate the source, and certainly such a weak concentration will not send a shark into an instant feeding frenzy. Still many species do rely heavily on their sense of smell to locate food.

continued on page 168

Below: *Nurse sharks are one of the few gregarious species of sharks that regularly rest on the bottom.* Opposite page: *An oceanic whitetip shark off Hawaii is accompanied by pilot fish. These fish swim near the shark's mouth waiting for scraps.*

SHARKS AND THEIR KIN

How Sharks Eat

Some sharks seem to have evolved with a single goal in mind. They eat. Much of their anatomy, from their powerful tails to their razor-sharp teeth, has evolved to improve their hunting skills. Many sharks, for example, have protrusible jaws. A special flexible ligament that connects the jaws to the skull allows the upper jaw to be pushed forward. When the shark opens its mouth to bite, the jaws and teeth extend forward for a more secure grip on the prey. Once the shark captures its food, the jaw and its teeth retract, or fold back toward the shark. Most large animals need to chew their food well as a part of the digestive process. Sharks, in contrast, have very large stomachs; their internal organs are able to handle huge chunks of food. Their mouths, then, do not have to be designed to aid digestion—their one and only role is to catch food.

Different sharks have different kinds of teeth, largely because they eat different foods. Great white sharks prey primarily on marine mammals, and they have large, triangular teeth with serrated edges. These teeth help great whites saw chunks of flesh from prey that is too large to swallow whole. In contrast, mako sharks eat fast-swimming fish such as tuna. Makos have spikelike teeth that snag or impale their smaller prey, which they then swallow whole. Many reef sharks have short, broad teeth with extremely sharp edges. These crushing teeth can break the shells of the turtles, crabs, lobsters, and mollusks that constitute their diet.

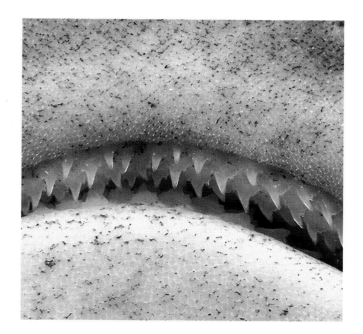

Opposite page: *The slender blue shark demonstrates its skill as a hunter.* This page, top: *The leopard shark's teeth are short and broad for crushing hard-shelled prey. Only the outer row of teeth is functional. As with all sharks, three to seven more rows lie behind the front row to move into position as front teeth are lost.* Above: *The sand tiger has long, spikelike teeth that add to its ferocious appearance.*

SHARKS AND THEIR KIN

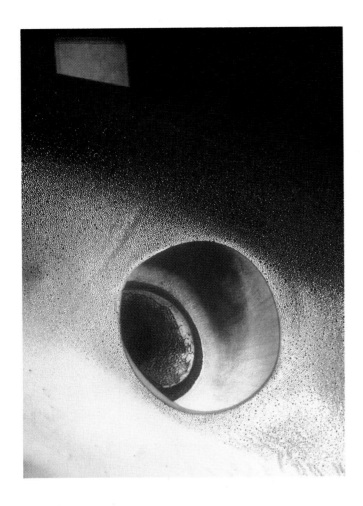

This page, top: *The nictating membrane begins to close over a blue shark's eye as it moves in to feed. The membrane protects the eye from being injured by prey as it thrashes in the shark's jaws.* Above: *A nurse shark moves after a school of porkfish and grunts. At this close range, the shark will rely on the electroreceptors in its snout to home in on its prey.* Opposite page: *The dogfish has one of the longest known life spans among sharks. This group also has one of the widest distributions, with species ranging from the Arctic to the Antarctic.*

continued from page 164

Like all fish, sharks have special nerves that run laterally along the sides of their bodies. These nerves, collectively referred to as the lateral line, help them feel things in the water around them. This nerve system is especially sensitive to vibrations, waves, and changes in pressure.

Sharks do have inner ears, but how much they rely upon hearing remains a point in question. Certainly hearing does play a part in the way sharks analyze their surroundings, but it is difficult for scientists to determine whether the ears or the lateral line do more to detect and analyze sound. Using the ears and lateral line together, sharks are especially adept at detecting low frequency vibrations. Injured fish that thrash about on the surface or that swim abnormally produce low frequency sound waves that arouse many species of sharks. Explosions that might accompany a sinking ship or an airplane that crashed at sea also produce low frequency sound waves.

Vision varies considerably from species to species. Most scientists believe sharks use sight to some degree to help analyze stimuli from a distance. Up close, within a few feet of their prey, sight is not as important as some other senses. The eyes of many species are designed to work best in low-light situations. Blue sharks, lemon sharks, and other species that live in bright, clear, shallow water probably rely upon vision more than sharks that live in deep, dark waters. Many sharks probably see their surroundings in hues of black and white, although laboratory studies have shown lemon sharks to respond to color.

SHARKS AND THEIR KIN

SHARKS AND THEIR KIN

SHARKS AND THEIR KIN

Egg Bearers, Live Bearers, and Intrauterine Cannibals

Sharks are an experiment in nature in that they give birth several different ways. Most bony fish produce a very large number of small eggs; males fertilize the eggs after the females lay them. Sharks have evolved a different reproductive strategy. Fertilization occurs internally, inside the female's body. Male sharks, rays, and skates have a pair of claspers, organs used during copulation to deliver sperm to the female. Sharks also produce far fewer eggs than most bony fish, but they do more to protect the developing young.

Blue sharks, oceanic whitetips, hammerheads, lemon sharks, and many others bear live young, while other species produce eggs. In some cases, the eggs develop and hatch inside the mother's body where they remain until they are more fully developed. In other species, females lay their eggs. Some of these species produce egg cases that hold the developing young after they are laid.

Some shark young are intrauterine cannibals. In some of the species where the young hatch inside the female, the developing sharks begin their intense battle for survival as soon as they hatch. Mako sharks, threshers, basking sharks, great whites, and others fight and devour their siblings and any unhatched eggs while still inside the mother's body. The surviving young gain added nutrition through their cannibalism so they are stronger and more developed when they emerge from the mother.

One of the most remarkable senses in sharks is their ability to detect electrical fields. They do this with special organs called the Ampullae of Lorenzini, which are small, gel-filled pits in the snout and other parts of the body. All living organisms create electrical fields around themselves, and the ability to detect these helps many sharks locate food sources hidden in crevices or buried in sand. Sharks can detect electrical fields that are far more faint than any other known animal can detect.

Opposite page: *A great white shark moves in like a torpedo. It is the only shark that feeds on adult marine mammals.* Above: *A baby swell shark, still in its egg case, is only about one and a half inches long. These sharks have a 13 month gestation period within the leathery cases.*

The ability to detect electrical fields also helps explain why sharks so often bite boat ladders, engine props, shark cages, and other metallic objects. Many people think this behavior proves that sharks are mindless, aggressive monsters. Actually, it's a perfectly natural and logical response to a certain stimulus. Any metal will create an electrical field when it is in salt water, and for millions of years, electrical fields have meant food to sharks. They are simply responding the way any hunter would when its senses say that prey is available.

Many sharks also have an inbuilt electromagnetic compass that is sensitive to the earth's magnetic fields. This ability helps explain how some species can make long open ocean migrations or hunt over great expanses of ocean without ever getting lost.

Above: *Great white sharks often attack shark cages, probably because they can sense an electric field in the metal rather than the human occupants. As the shark bites, its jaws protrude out of its mouth for a better grip.* Opposite page: *Oceanic whitetip sharks feed on the carcasses of dead whales, among other things.*

THE DISTRIBUTION OF SHARKS

Sharks inhabit all oceans of the world, from the deepest parts of the coldest seas to the shallowest tropical regions. The exact distribution by species depends largely on water temperature, so sharks are generally put into three major groups. They are the sharks of tropical waters, temperate waters, and cold waters. Each group breaks down further into active swimmers and bottom dwellers. As a general rule, the active swimmers tend to be larger, and they often have a larger territory than the bottom-dwelling varieties.

The tropical open ocean sharks include whale sharks, mako sharks, oceanic whitetips, and thresher sharks. Active swimmers in tropical reef communities include tiger sharks, hammerheads, gray reef sharks, bull sharks, lemon sharks, blacktips, silky sharks, and many more species. Bull sharks will sometimes even visit fresh water rivers and lakes. Carpetsharks or wobbegongs, banded catsharks, nurse sharks, and whitetip reef sharks are among the most common bottom dwellers in tropical reef communities.

The active swimming sharks of temperate waters tend to migrate over the course of a year. These species generally move toward the equator during winter months and away from the equator in the summer. The species with the widest distribution include basking sharks, great white sharks, blue sharks, sand tigers, and threshers. Of these, blues probably cover the greatest distances on a regular basis. Individuals caught near

SHARKS AND THEIR KIN

This page, top: *Blue sharks off the coast of California are easily attracted by chumming, a technique to lure ocean life by throwing pieces of fish into the water from a boat.* Above: *A six-gilled shark is caught in the light of a submersible. Most sharks have five gills; only a few have six or seven.* Opposite page: *Great white sharks can be tempted part way out of the water by a chunk of meat.*

New York have traveled as far as Spain, and specimens tagged off the coast of England have made their way to the waters of Brazil. Horn sharks, angel sharks, some catsharks, and some sawsharks are bottom-dwelling temperate species. As with most bottom dwellers, they are comparatively small and quite docile.

Many species of sharks also inhabit cold waters. Among the larger active cold water species are the six-gilled and seven-gilled sharks, goblin shark, Portuguese shark, and Greenland or sleeper shark. Greenland sharks sometimes venture under polar ice floes, and the Portuguese shark can dive to almost 5,000 feet below sea level.

THE NATURAL HISTORY OF SOME REPRESENTATIVE SPECIES

Perhaps the most distinctive thing about the shark family is its great diversity. Each species has its own adaptations, its own lifestyle, its own role in nature. Of the 350 known species, though, some are noteworthy because of their behavior, size, natural history, or because they are typical of a group.

Great White Sharks

Although much remains uncertain about their natural lives, great whites are probably the most famous sharks. Certainly they are the most feared. To scientists, they are *Carcharodon carcharias*, but to many tourists at the beaches they are white death. A little notoriety, a starring role, and some

SHARKS AND THEIR KIN

publicity mixed with truths and half-truths have created a reputation that is far more mythical than real. Their size and power make them a threat to almost any animal in their domain, but they are not indiscriminate feeders that attack anything that moves, as many people think.

Though great whites live in all oceans, people most often encounter them in temperate seas in the shallow waters over continental shelves. The largest documented specimen was 21 feet long and weighed just over 7,000 pounds. Typical adults range from 12 to 16 feet long and weigh from 1,800 to 3,500 pounds. Newborn great white sharks are about 3 to 4 feet long.

The teeth of great white sharks are triangular with serrated edges that look similar to the blade of a saw. Like other sharks, the great white has many rows of replacement teeth. When a tooth is lost, a new one simply moves forward to take its place.

During the first few years of their lives, great whites feed upon a variety of rays, flatfish, and other animals that live in sandy areas or in nearby reef communities. As they become larger, their diet shifts toward marine mammals. Adult great whites hunt sea lions, seals, and dolphins. They also sometimes feed on other sharks, salmon, sturgeon, hake, rockfish, lingcod, mackerel, tuna, and other fish.

SHARKS AND THEIR KIN

The Maneaters

Sharks and humans evolved on earth along separate paths involving different environments. The feeding patterns that make sharks successful predators developed millions of years before people ever appeared on earth. Sharks hunt a wide variety of fish, marine mammals, crustaceans, mollusks, echinoderms, and some other food sources. Human beings are definitely not part of their natural diets.

Still, sharks do sometimes attack humans. About 300 documented attacks occur every year, and some cases clearly involve intentional feeding. The low frequency sounds created by explosions attract some sharks. As a result, sharks have killed victims of air and sea disasters, many of whom were injured and bleeding. Blue sharks, oceanic whitetips, tiger sharks, and a variety of tropical reef sharks are the primary culprits.

In most attacks, though, the shark is responding to human provocation. That usually means underwater photographers using bait carelessly, spearfishers trailing their take through the water, or divers unknowingly violating the domain of a territorial shark. Sometimes a diver or snorkeler will cause an attack by foolishly trying to ride or handle a normally docile whitetip reef shark, nurse shark, or angel shark.

A shark will also sometimes mistake a human for its regular food source. For example, when great white sharks attack surfers, they probably think they are attacking a sea lion or seal. From below, the silhouette of a person resting on a surf board strongly resemble a sea lion or seal resting on the surface.

The species involved in documented attacks include great white sharks, shortfin mako sharks, blue sharks, lemon sharks, gray reef sharks, tiger sharks, bull sharks, three species of hammerheads, dusky sharks, and bronze whalers. These species are not really that much more aggressive than other sharks, but they do populate areas where humans spend time in the ocean.

Opposite page: *Although they have a vicious reputation as maneaters, great whites probably most often attack humans by mistake. This theory is supported by the fact that most attacks entail only a single bite by the shark. After an initial taste, the shark often leaves the victim alone, perhaps realizing that it has gone after the wrong prey. Unfortunately, even a single bite from such a creature can be fatal.* Left: *Makos, too, will occasionally stray from their normal diet of tuna and other open ocean fish and attack humans.*

SHARKS AND THEIR KIN

Scalloped Hammerheads

Hammerheads are relatively modern sharks, having evolved some 120 million years ago. There are nine different species, and their most striking feature is, of course, their hammer-shaped head. Their unusual design serves two important functions. First, the head's shape provides added lift at the front end of the body, similar to the way wings provide lift in airplanes. In addition, a wider distribution of sensory organs may enable hammerheads to locate prey and predators more quickly.

Scalloped hammerheads demonstrate a rather unusual grouping or schooling behavior. As a rule, most sharks never form large groups except when attracted by bait or when mating. Scalloped hammerheads, however, often gather by the hundreds in many parts of the tropical Pacific. Individuals within the group frequently show distinctive swimming patterns that might be some form of communication. Despite more than a decade of study, scientists are still uncertain why these sharks school.

Blue Sharks

Named for their iridescent color, blue sharks are among the most beautiful of all shark species. Their magnificent color is most obvious on sunny days when the sharks swim near the surface. Their skin sparkles in stunning hues of blue as the sunlight dances over their backs. Blue sharks are easy to recognize by their color, long slender bodies, and comparatively long pectoral fins. Like many open ocean predators, blue sharks are countershaded; they have a dark top side and a lightly colored underside.

Blue sharks inhabit the surface waters of the open ocean in all temperate and some tropical seas. They are typically six to seven feet long and weigh approximately 65 to 110 pounds. Though reported to 18 feet, the largest blue shark ever documented was just under 13 feet long. Blue sharks prey on squid and small schooling fish such as anchovies, herring, and mackerel.

Opposite page: *The Sea of Cortez is one location where scalloped hammerhead sharks gather en mass—sometimes by the hundreds. No one has yet discovered why.* Above: *A blue shark cruises its stalking grounds, the surface waters of the open sea, in search of prey.*

Gray Reef Sharks and Galápagos Sharks

Gray reef sharks and Galápagos sharks are typical of most reef sharks. While these sharks do not live in schools, they tend to congregate in reef communities. These sharks often participate in wild feeding frenzies.

This page, top: *Gray reef sharks are one of the few species that are fiercely territorial.* Above: *A Galápagos shark is shadowed by blue runners, probably waiting for the spoils of the shark's next feast.* Opposite page: *An array of dots and spots decorate the whale shark's upper body. Most large sharks lack such markings.*

Gray reef sharks are also well known for their exaggerated posturing, radically hunching their back and tucking their pectoral fins before they strike. These threat displays do not precede all attacks by gray reef sharks, but once the sharks display, they almost always attack.

The average Galápagos shark is six or seven feet long, about the size of an average gray reef shark. Both species are capable of rapid bursts of speed, and they have excellent maneuverability, a necessary feature for hunting in the tight confines of the reef.

Whale Sharks

Inhabiting all tropical seas, whale sharks are the world's largest fish. They attain a size of 40 feet and 30,000 to 40,000 pounds, and many specialists suspect they get even larger. The brownish or black upper portion of a whale shark is covered with large yellow spots, while the underside is off-white in color. A series of pronounced ridges run the length of the body. The mouth of a 40 to 50 foot long whale shark is between six and ten feet wide and holds approximately 5,000 tiny teeth.

Like most of the oceans' largest creatures, whale sharks are filter feeders. They consume huge quantities of small crustaceans and some small fish in the two to six pound range. When feeding, whale sharks open their mouths and pass an enormous volume of water through the mouth and gills. The prey gets caught on the gill rakers. The sharks close their mouths and tilt their heads up, forcing the water out, and then they swallow the captured food.

SHARKS AND THEIR KIN

SHARKS AND THEIR KIN

SHARKS AND THEIR KIN

Countershading: Coloration for an Advantage

Many sharks rely heavily on camouflage to avoid potential predators and capture prey. Bottom dwellers, for example, can often rest unnoticed among rocks or on the sandy bottom because of their coloration. Active-swimming blue sharks, makos, gray reef sharks, and great whites use a different sort of camouflage called countershading. These sharks are dark on the top portion of their bodies and lightly colored underneath.

Countershading works because sunlight does not penetrate deep water. The waters near the surface contain more light and are lightly hued; at greater depths there is less light and the water is dark blue to black. So if a sea lion is swimming above a great white shark and looks down, it will see the dark upper portion of the shark against the dark background of deep water. On the other hand, if the shark is swimming above the sea lion, the shark's light underbelly will blend into the light-colored surface water.

It is difficult to believe that a 16-foot, 3,000-pound shark can hide no matter what its color, but countershading is remarkably effective. Divers often see huge sharks completely disappear from view 30 or 40 feet away even though water visibility is at 100 feet or more.

Horn Sharks and Angel Sharks

Although horn sharks and angel sharks do not look at all alike, they are typical of many bottom-dwelling sharks. Horn sharks look very much like catfish. Found in the kelp forest communities of western North America, they reach a maximum size of about three feet. They prey upon a variety of crustaceans, mollusks, echinoderms, and small fish.

Angel sharks have extremely flattened bodies. Many divers mistake them for rays. They are lithe, graceful swimmers, but they spend most of their time lying on the seafloor. Their body shape and coloration blend seamlessly with the sandy bottom. Prey will often swim unaware within inches of an angel shark's mouth.

Opposite page: *The bottom-dwelling horn shark is named for the sharp spine in its dorsal fin that could lodge in the throat or mouth of predators.* Above: *From the front, a flattened, wide angel shark looks more like a ray than a shark. A side view, however, reveals a more typical shark body and tail.*

SHARKS AND THEIR KIN

RAYS AND SKATES

This page, top: *A manta ray glides gracefully through the water, filter-feeding as it goes.* Above: *The sawfish, a type of skate, uses its long, serrated snout, or rostrum, to uncover prey buried in the sand. It may also use its saw to stun or injure prey.* Opposite page: *Eagle rays seem to fly through the water with an ease and grace few other ocean dwellers can achieve.*

Skates and rays appear in about 470 different species. Skates look very similar to rays and to some bottom-dwelling sharks. The simplest way to distinguish between the three is to check the gills and tails. A shark's gill slits are always on the side of the head, while the gills of rays and skates are on the underside. Also, rays have thin, whiplike tails, and sharks and skates have fleshier, more prominent tails.

Rays thrive throughout tropical and temperate seas. Some species such as manta rays and mobula rays are pelagic creatures, meaning they prefer the waters of the open sea. Others such as stingrays, bat rays, and spotted eagle rays frequent reef communities. Manta rays and mobula rays rarely, if ever, approach the seafloor, while many species of stingrays live their entire lives near the bottom.

Superb swimmers, eagle rays often cruise along the seaward edge of a reef. From there, they can enter the reef zone to hunt crustaceans and mollusks or quickly retreat to the open sea where they can outswim their predators.

Surely the most impressive of all the rays is the manta ray. People once viewed these gentle giants as dangerous monsters of the deep. The manta's cephalic lobes, hornlike appendages protruding from the head, prompted some to call it the devil fish. The lobes act as a funnel, directing plankton and small fish toward the manta's mouth.

SHARKS AND THEIR KIN

185

SHARKS AND THEIR KIN

SHARKS AND THEIR KIN

Manta rays can grow to 20 feet across, and at that size they weigh at least 3,000 pounds. The huge, flattened pectoral fins provide considerable thrust and can move independently, making mantas excellent swimmers. Divers rarely get a close-up look at mantas. The rays can easily outdistance humans with only the slightest movement of their fins.

Unlike manta rays, most stingrays live close to the seafloor. Stingrays have one or more knifelike barbs or spines near the base of their tails. The barbs are often large, but they are strictly defensive weapons. Stingrays have very flexible bodies, and they can place the barbs with great accuracy. The barbs usually carry bacteria that can lead to nasty infections, and they also contain a powerful toxin.

Stingrays' mouths rest on the underside of the head. As they rest on the bottom, they flap their large pectoral fins to stir up the sediment and uncover a meal of crustaceans or mollusks.

Most rays are excellent swimmers, but of course there are exceptions. Electric rays, also called torpedo rays, are sluggish bottom dwellers. They have large specialized glands that produce electricity for fighting predators and stunning prey. The electrical intensity varies from species to species, but the shocks can cause considerable pain in humans. Creating the electricity requires much energy, and the rays use it sparingly.

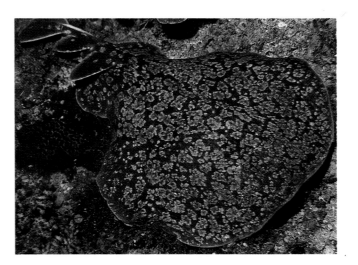

Opposite page: *The mouth and two sets of gill slits are located on the underside of a southern stingray.* This page, top: *An electric, or torpedo, ray can deliver a nasty shock to anyone who dares to touch it.* Above: *Cownosed rays group in massive schools off the Galápagos islands.*

MARINE MAMMALS

Without question, humans hold marine mammals in high regard. Many species show ingenuity and strength in overcoming the ocean's rigors. We admire their beauty and grace and feel a certain kinship with these other mammals who in many ways rule their environment.

Marine mammals include whales, dolphins, seals, sea lions, walruses, sea otters, and manatees. These species evolved from land mammals, and they retain many of the distinguishing mammalian traits. They are air-breathing, warm-blooded vertebrates that bear live young, nurse their offspring, and have hair or fur at some point in their development.

The ocean presents some unique survival problems for marine mammals. Water absorbs body heat more rapidly than air does, and they must struggle to maintain their body temperature. Some marine mammals solve this problem with blubber, thick insulating fat that keeps the cold out and the heat in. The fat may also provide energy reserves when food is scarce. Seals and sea lions have thick fur to help keep themselves warm. Sea otters grow dense fur and renew their energy by eating huge amounts of food.

Walrus bulls with their long, white tusks (opposite page), *a sleek sea otter floating on its back* (above, left), *and the impressive killer whale* (above) *have more in common than a fondness for water. As mammals, they are among the most highly developed creatures in their habitats, and they must constantly struggle to maintain their body temperatures and find access to life-sustaining air.*

MARINE MAMMALS

This page, top: *A group of California sea lions displays the playful social behavior they are known and loved for.* Above: *Although an excellent swimmer, this harbor seal hauled itself onto a rock for a little rest.* Opposite page: *These seals and gulls have both come across the same food source, probably a large school of small fish.*

Marine mammals have also developed some unique reproductive strategies. Many of them have very long gestation periods, so their newborns are well developed and have a better chance to survive. Many aquatic mammals sharply cut their food intake during their mating seasons, too. The adults can spend more time guarding, nursing, and training their young. Some species also eat less during migration to hasten the trip between feeding areas and mating grounds.

PINNIPEDS

Seals, sea lions, fur seals, and walruses are all pinnipeds. The term comes from Latin words meaning "fin footed." In these species, the limbs have evolved into flippers, an obvious advantage in their aquatic lifestyle.

As divers, pinnipeds benefit from many special adaptations. Their streamlined shape reduces water friction and makes swimming easier. Their bodies are also very flexible, allowing them to maneuver quickly and gracefully. During their dives, the heartbeat and overall metabolism slow to save oxygen.

Pinnipeds spend most of their time in the sea, but unlike dolphins and whales, they also have strong ties to land. They often come ashore to bask in the sun and rest, and they also gather on land or ice to breed and bear their young.

Some pinnipeds have developed a reproductive strategy called delayed implantation. In some species, males and females

MARINE MAMMALS

MARINE MAMMALS

gather for birthing and mating at the same time. Females often breed only days after bearing their young. The resulting fertilized eggs do not settle in the uterine wall where they can develop until some time later. The fetus does not draw energy from its mother for several months, so she can nurse her newborns and regain her strength unhindered by the pregnancy.

Sea Lions and Seals

Perhaps the most surprising thing about seals and sea lions is that they may not be so closely related as people think. Scientists have yet to agree on their exact development, but there is some evidence that seals and sea lions evolved from different ancestors at different points in history. The most obvious difference between these two pinnipeds is that sea lions have external ears and seals do not.

Most seals and sea lions inhabit temperate and polar waters. Different species of sea lions and seals gather together both in the water and on land. San Miguel Island off Southern California boasts the largest mixed gathering of pinnipeds in the world.

There are five species of sea lions. Steller sea lions are the largest of the eared seals. Restricted to the northern hemisphere, males weigh as much as a ton.

California sea lions range all the way from British Columbia to the Galápagos Islands near Ecuador. Easily trained, they often star at marine parks around the world. In the wilderness, they are very social, as most pinnipeds are.

Opposite page: *The California sea lion's great speed and agility will ensure that some of these fish end up as a meal.* This page, top: *Gregarious California sea lions often swim and fish in large groups.* Above: *California sea lions, along with the other eared seals, have short, furled outer ears that distinguish them from the earless, or true, seals.*

MARINE MAMMALS

This page, top: *A leopard seal has flippers with five webbed digits, and it lacks external ears. Although it looks harmless, this species is known to be quite aggressive.* Above: *Sea lions have hauled themselves onto a rocky Alaskan shore to get some sun, a common sight during breeding seasons.* Opposite page: *Off the coast of Australia, a blonde sea lion swims toward the surface for a breath of air after a long dive.*

Pups and females are very interactive. Adult males, called bulls, compete vigorously with one another to establish and defend territories. The competitions include bluffing and displays, but they also lead to all out fights—usually chest-to-chest pushing and biting.

The bulls' crested skulls distinguish them from the females. Males breed with all the females found in their territory, but they make no effort to keep the females from mating in other areas.

The white markings and graceful antics of Australian, or blonde, sea lions make them among the most stunning pinnipeds. Nonmigratory, they occur only in South Australia. Like many pinnipeds, they are a favorite food source for great white sharks. South American sea lions and Hooker's sea lions of New Zealand are the other southern hemisphere species.

Most pinnipeds feed largely upon fish, mollusks, and crustaceans, but the leopard seal is a notable exception. Prowling the waters of the Antarctic, they also feed on penguins and other seals. Female leopard seals are larger than the males. Photos of this species are rather rare, due as much to the animal's aggressive temper as to its inhospitable environment.

Weddell seals are another of the Antarctic seals. They live farther south than any other mammal. They spend much of their time in the water, diving under the thick polar ice. Using specially modified canine teeth, Weddell seals bore holes through the ice to breathe.

MARINE MAMMALS

MARINE MAMMALS

This page, top: *Two elephant seal bulls fight for a position on this South Georgia Island beach where females will soon come to mate. The males establish a rigid social hierarchy during breeding season. These two young bulls lack the overgrown nose that male elephant seals are known for; the nose becomes prominent later in life and serves as a display during competition. These two will probably not mate this season. Older, larger males will drive them from the beach.* Above: *Northern fur seals are the most oceanic fur seals in the northern Pacific, coming ashore primarily during the breeding season.* Opposite page: *A young elephant seal pup sits alone on San Benitos Island near Mexico.*

How Marine Mammals Avoid the Bends

Many marine mammals dive to phenomenal depths. Sperm whales dive 2,500 feet, perhaps even deeper, in search of food, and northern elephant seals can reach depths of 2,900 feet. When diving this deep, they are exposed to enormous pressure, pressure that can cause the mammals known as human beings to get the bends.

The bends occur as a result of the way increased pressure affects gases in the body. As water depth increases, so does the pressure of water. Human divers breathe compressed air at a pressure equal to that of the water surrounding them. The increase in pressure forces gases in this air, particularly nitrogen, to dissolve in a diver's bloodstream and tissues. If the diver ascends too quickly, causing a rapid reduction in pressure, the nitrogen gas may form bubbles. These bubbles can become trapped in the diver's tissues, causing intense pain, tissue damage, and in extreme cases, paralysis or death.

As air breathers marine mammals have evolved ways to avoid the bends. Most importantly, they do not breathe compressed air as humans do when they dive. They fill their lungs at the surface, where the surrounding pressure is less, and do not intake any air during their dives. This greatly reduces the amount of nitrogen that dissolves in their bloodstreams. Also, their lungs compress when they dive so that less gas exchanges between their lungs and their bodies. By reducing gas exchange, they lessen the amount of nitrogen in the bloodstream and avoid the risk of air bubbles being trapped.

Southern elephant seals, which live in the southern Atlantic and Indian oceans, are the largest pinnipeds. Full grown males can be 15 feet and 5,000 pounds. A close relative, the northern elephant seal, reaches a top weight of 4,000 pounds.

Ribbon seals are one of the most attractive pinnipeds. Inhabiting the Bering, Chukchi, and Okhotsk seas, these particularly agile seals are born white. After several molts they develop a handsome banded pattern that leads to their common name.

The seals of the Arctic Circle include ringed seals, hooded seals, bearded seals, and harp seals. Harbor seals actually range throughout the temperate and polar waters of the north. They have a chunky shape and lightly colored coat covered with dark spots. They are one of the smaller species, and they often show curiosity or interest in snorkelers and divers.

Harp seals have distinctive harp-shaped patterns on their backs. They have achieved worldwide attention recently due to heavy pressure from hunters in Greenland and Newfoundland who value the pups' white fur. New restrictions limit the hunting of infants, but much controversy remains over their future.

MARINE MAMMALS

Walruses

Found only in the Arctic, walruses are famous for their impressive tusks, which are actually canine teeth. The tusks are long and straight in the males and shorter and curved in the females. The tusks grow throughout their lives. In older males, they can exceed three feet.

Walruses may use their tusks to dig the sea bed for worms, mollusks, and fish. They also use the tusks to pull and brace themselves as they travel the icy arctic landscape. Mature males battle each other for territory and defend against predators such as polar bears and killer whales with their tusks.

Walruses are among the largest pinnipeds found in the Arctic region. Adult males commonly reach 11 feet and 3,500 pounds. Walruses live almost exclusively near coasts on pack ice and ice edges. During the bitter cold winter months, they migrate south, but they always remain around the ice floes.

Walruses are highly gregarious animals, and they commonly gather in the thousands. Like other pinnipeds, they follow very specific mating patterns. They breed every other year, and 15 months later the female bears a single pup. A new-born walrus is about four feet long and weighs close to 110 pounds. The female nurses her young for up to two years.

Opposite page: *A huge mass of sleeping walruses litters a remote rocky shore. Their distinctive tusks can weigh up to 12 pounds each.* This page, top: *On Round Island in Alaska, a couple of walrus bulls take a moment to have a playful fight.* Above: *Adult male walruses may lack the characteristic short, rusty brown fur.*

MARINE MAMMALS

SEA OTTERS

Members of the weasel family, sea otters thrive along North America's western coast where their playful antics keep observers entertained for hours. Their reddish brown to black fur is dense, soft, and very fine. They venture onto land where they can travel several hundred yards from shore, but they spend most of their lives at sea. Sea otters are the only marine mammals that lack blubber to keep them warm. Yet they live in cold water, and like all other active mammals, they must maintain their body temperature.

Above: *Humans aren't the only mammals that use tools. Sea otters use rocks to break open mollusks and crustaceans at the surface and to pry or knock them loose underwater.* Opposite page: *Sea otters' dense fur does much to keep them warm, but they must meticulously groom themselves to maintain their coats.*

To solve their problem, sea otters eat, and eat, and eat. Juveniles consume up to 35 percent of their own body weight in a day's feeding. Mature adults devour up to 25 percent of their weight daily in the form of sea urchins, lobsters, clams, crabs, squid, scallops, tubeworms, sea stars, snails, and more. Male sea otters reach a length of almost five feet and can weigh as much as 85 pounds; an adult male can consume 5,000 pounds of food each year. Females are smaller than males, reaching a size of four feet and 60 pounds when full grown.

Sea otters are one of only a few animals that definitely use tools. They frequently lay on their backs at the surface and use large rocks to crack open clams, mussels, and other hard-shelled prey items. Below the water, sea otters wield rocks like hammers as they knock scallops and mussels off the rocks.

Otters are excellent swimmers and divers. They prefer to hunt in shallow water, to depths of 60 or 70 feet, but they sometimes venture deeper. Their hunting dives commonly last a minute or more, but they can stay submerged as long as four minutes.

Sea otters once ranged from the shores of Mexico's Baja Peninsula north to Alaska, eastward across the Bering Sea, and south to Japan. In the late 1700's, hunters began to pursue them for their pelts. After 170 years of hunting, their populations neared extinction. Today sea otters are making a comeback, but it is an effort full of controversy. Many fishermen oppose the sea otters' success because they consume the shellfish that support the fishermen.

MARINE MAMMALS

MARINE MAMMALS

MANATEES

Manatees live in shallow, coastal waters in tropical regions; they never venture onto land and only rarely enter the deep ocean. Manatees, commonly called sea cows, belong to the order Sirena, the sirens, because legend says that ancient sailors mistook them for mermaids.

The order includes four separate species. Three are manatees that live in the Atlantic basin—the West Indian, African, and Amazon manatees. The fourth is the dugong, which lives in the Indo-Pacific. In North America, manatees inhabit several Florida river systems, and they thrive in the coastal estuaries of many Central American countries, especially Belize.

Manatees are herbivorous. Adults consume up to 100 pounds of plant life a day to maintain their weight. Their rubbery lips are covered with bristles, and they are well adapted for devouring huge amounts of sea grasses. As herbivores, they have strong molars for grinding and chewing. They lack cutting teeth like the incisors that humans and many other predators have.

As is so often the case, the major threat to manatees comes from humans. During the 1700's, hunters drove the largest manatee, the Stellar sea cow of the Bering Sea, to extinction in a 27-year period. Stellar sea cows reached a length of 25 feet and weighed up to 9,000 pounds. Today, loss of habitat due to commercial development and careless boaters who run over manatees are the primary threats.

Opposite page: *Manatees are slow-moving animals that generally swim near the surface. Many are hit by boats every year.* This page, top: *In the winter, Florida manatees swim up to freshwater springs where the water is warmer than in the ocean. Here, they perform a service to humans and other animals by feasting on the prolific water hyacinth that would otherwise choke many waterways.* Above: *A manatee calf approaches its mother to nurse.*

MARINE MAMMALS

CETACEANS

The order Cetacea consists of dolphins and whales. Collectively they are the oldest and best-adapted group of aquatic mammals. Their forelimbs have evolved into flippers joined to the body at the shoulder, and they have completely lost their hindlimbs. Their large tail flukes and dorsal fin greatly assist them in their aquatic lifestyle, as do the blowholes atop their heads. The blowholes serve as an airway to the lungs.

There are two major categories or suborders of dolphins and whales. Baleen or filter-feeding whales are in the suborder Mysticeti, and toothed whales are in the suborder Odontoceti. The two groups differ in several major ways, but the biggest difference is that the baleen whales lack teeth after birth. They have instead long horny plates called baleen that look somewhat like the fibers of an oversized toothbrush. The baleen filters great quantities of plankton such as krill, copepods, isopods, and some small fish from the water when baleen whales feed.

As the group name implies, toothed whales such as killer whales, sperm whales, and all dolphins and porpoises have teeth. Their diet is quite different from that of baleen whales. Toothed whales feed on fish, squid, and octopi; killer whales even eat seals, sea lions, and other whales.

A bottlenose dolphin (opposite page), *a humpback whale* (this page, top), *and a killer whale* (above) *breach the surface, each in its own spectacular fashion. Breaching is common behavior among cetaceans, but the reasons for it remain unclear. It may help rid them of parasites or allow them to navigate using coastal landmarks, or it may be something they do simply for the joy of it.*

Baleen Whales

The baleen or filter-feeding whales are often referred to as the great whales due to the enormous size of many species. The blue whale is not only the largest filter-feeding whale, but it is also the largest animal living on earth today. Blue whales can grow to an incredible 110 feet long and weigh as much as 120 tons. Fin whales, humpbacks, sei whales, gray whales, and other baleen whales also reach tremendous sizes.

Minke whale and pygmy right whales are also filter feeders. As adults these species are much smaller than some of their close filter-feeding relatives. In fact, some toothed whales such as killer whales and

continued on page 210

Above: *A whale's distinctive spout as it exhales comes from the moisture in its warm breath and the sea water that pools over its blowhole.* Opposite page: *A California gray whale spyhops, lifting its head out of the water far enough to take a look around.*

The Migration of the California Gray Whale

Every year, huge numbers of California gray whales migrate between the Bering Sea off Alaska to a series of protected lagoons along the Pacific coast of Mexico's Baja Peninsula. The 10,000-mile round trip is the longest mammal migration known to science.

The migration centers on the whales' calving, breeding, and feeding cycle. During the summer and early fall, gray whales feed in the prolific waters of the Bering Sea. Beginning in late fall and continuing through winter, they begin the southward part of their journey, which takes them along the coast of western North America. Pregnant females are usually the first to depart northern waters and the first to reach the lagoons. The males soon follow.

The females give birth in the lagoons soon after their arrival. They nurse their young for about seven months. The adults also mate during the winter, with the pregnancies lasting about 13 months. In the early spring, the herds return north, and this portion of the journey takes the whales farther out to sea.

Animals undergo lengthy migrations for a variety of reasons. With the gray whales, the journey allows them to avoid the bitter winter conditions and limited food supply of the Arctic winter. It is also possible that the warmer, more secure lagoon waters give the new offspring a greater chance to survive.

MARINE MAMMALS

MARINE MAMMALS

Songs of the Humpback Whales

For centuries, mariners told stories of eerie, mournful dirges rising from the oceans' depths. Scientists long doubted that humpback whales were the source of these cries. The animals lack vocal chords, making it unlikely that they would be able to produce such intricate sounds. In the early 1950's, however, underwater microphones recorded the humpbacks' now famous songs.

The songs of humpback whales are the longest and most complex sound sequences produced by any whale. A variety of other baleen whales also produce a series of low moans and other sounds. The low-frequency sounds can travel up to a hundred miles in the water.

Scientists continue to debate exactly how and why the whales sing. They believe that only the male humpbacks sing and that the songs are a form of communication with other whales, but they know little beyond that. The songs could be a means of attracting or communicating with a mate, threatening other males, defining territory, or sharing knowledge about food sources.

The songs typically consist of moans, groans, sighs, roars, and high-pitched chirps and peeps. A given sequence might last as long as 35 minutes and then repeat in its entirety with no variation at all. Whales often repeat precise sequences for hours on end.

In any one area, all the humpbacks sing the same song, but the songs change a great deal over the course of a given year. The songs change gradually, and the changes occur simultaneously in all the whales in a given area. Populations in different geographical areas such as the Caribbean, Hawaii, and Australia sing songs that differ significantly from one another.

Opposite page: *In the waters off Hawaii, a humpback whale puts on an impressive display. The grooves on the underside of this huge creature's jaw expand when the whale opens its mouth to filter feed.* Left: *A mother, or cow, humpback shelters a calf under her pectoral fin for protection. Many whales live in close family groups and will go to great lengths to protect their offspring and their mates.*

MARINE MAMMALS

This page, top: *A humpback whale lobtails, showing off flukes that are as distinctive as an individual name tag. Scientists use the unique flukes to catalog wild whales and track their movements.* Above: *The long, white pectoral fins of a humpback whale can be a quarter to a third the length of the animal.* Opposite page: *Blue whales, growing to nearly 100 feet, are the largest of all whales. They are, in fact, the largest living creatures on earth.*

continued from page 206

sperm whales grow to larger sizes than the minkes and pygmy right whales.

Baleen whales usually migrate annually from their polar or subpolar feeding grounds to their calving and breeding areas in temperate and tropical seas. These whales usually spend the warmer months in their feeding areas before moving toward the equator during polar winters.

Because plankton occurs densely in shallow waters, baleen whales spend much of their lives in the top 150 feet of the water. However, they are very capable swimmers, and their dives often last as long as 20 minutes.

All 11 known species of filter-feeding whales use their baleen as the primary way to catch food. Some species, though, also use some other techniques. Humpback whales, for example, feed on small schooling fish by using a technique called bubble netting. They force the fish into tighter and tighter schools by getting below them and exhaling air bubbles that surround and frighten the prey. The whales then rush straight up with mouths agape and capture as many fish as they can.

Right whales got their name because they were the best, or the right, whale for whalers to hunt. They travel slowly, are relatively easy to find and kill, float when they are dead, and have great commercial value. Sadly, their fate remains in question today

MARINE MAMMALS

Diving with the Great Whales

For many scuba divers, diving with the great whales ranks as an ultimate experience. Blue whales, humpbacks, and other baleen whales have body parts—the head, tail, flippers—that are much larger than a diver. The imposing presence of such a creature is an indescribable experience.

Many whales, even adults of the larger species, are extremely wary of humans. They often go out of their way to avoid people. Occasionally, though, whales will show some tolerance and not flee from human divers.

Divers who have been eye to eye with a great whale almost always speak of the meeting in reverent terms. They marvel at the grace and superb body control the animals show. Many divers have reported that whales repeatedly maneuvered around them, often coming within inches without ever touching.

due to intense hunting pressure. An estimated stock of 100,000 animals only a few hundred years ago has dwindled to a population of no more than 4,000.

The six species of rorqual or finback whales bear throat pleats or grooves that no other whales have. Working like the folds of an accordion, these pleats run lengthwise and let the throat expand greatly when the whales feed. Blue whales can easily filter several tons of water in a single mouthful.

Opposite page, this page top, and above: *These divers are dwarfed by the right whales they have been lucky enough to encounter. Although these baleen whales are filter feeders and will not bite a diver, humans in the water must be very careful around these animals. Whales almost never show any unprovoked aggression toward humans, but their size and power mean that even a slight, inadvertent swipe with a tail or fin can cause serious injury. Still, most divers probably feel that an up-close look at these majestic creatures is worth the risk.*

MARINE MAMMALS

Toothed Whales

Toothed whales include a number of species of whales and those species commonly called dolphins and porpoises. While this surprises many people, it is taxonomically correct. The whales are generally larger than the dolphins and porpoises, but these animals still share many traits.

Having teeth enables this group to pursue individual prey rather than filter feeding like the baleen whales. This means that they can populate a wider range of areas where plankton is not abundant.

Over time, toothed whales have diverged and specialized. Several species live exclusively in fresh water, others are strictly marine, and still other species can live in both fresh and salt water. This diversity also means that several species can live near one another. Different species tend to pursue different foods, so they can inhabit the same region without competing too heavily for food.

Reaching a size of 60 feet and 45 tons, sperm whales are the largest toothed whales. Their name comes from the spermaceti oil that made them a favorite target of whalers. Sperm whales feed on a variety of foods, but deep-sea squid are a major part of their diet. The whales dive as far as 2,500 feet, perhaps even deeper, as they search for food. The longest recorded dive of a sperm whale lasted two hours and 18 minutes.

Killer whales are among the more publicized toothed whales. They often star in

Opposite page: *The bulge, or melon, on the top of a beluga's head is filled with fatty tissue saturated with oil. The melon may be a part of this toothed whale's echolocation system.* This page, top: *Snowy white beluga whales, residents of the far northern Pacific, are not well studied in the wild because they rarely venture into more hospitable waters.* Above: *Like all whales, belugas are remarkably lithe and graceful despite their great size.*

marine park shows around the world. Killer whales are also known as the species with the towering dorsal fins. Actually, only the males have the huge dorsals, which can rise five or six feet high off their backs. Like most toothed whales, killer whales are social animals, and they often live in large pods. Their group behavior may be matriarchal, meaning that a female assumes the leading role in each pod.

The killer whales of Johnstone Strait near Vancouver, Canada, are probably the world's most studied killer whales. There are two local populations called resident pods; other killer whales called transients visit but do not live in the area year round. The two resident pods rarely mix, but both groups feed primarily on salmon, and they are highly vocal. The transient whales prefer marine mammals, mostly sea lions, seals, and Dall's porpoises. They are often silent, probably because the mammals they feed on have excellent hearing.

Narwhals are another interesting species of toothed whales. Their most notable feature is a highly developed tooth, or spiral tusk, that extends out from a male's upper jaw. Adult narwhals reach about 14 feet, and the tusk adds another six to eight feet to their length. Narwhals are an Arctic species. They feed on squid, fish, and a variety of crustaceans.

People know many species of toothed whales as dolphins and porpoises. They are typically smaller than the species called whales, and that is the only real difference between the two groups.

continued on page 221

Below: *In a tense, dramatic moment, a killer whale pursues its prey to water's edge. The meal could cost this whale its life if it becomes stranded on the sand.* Opposite page: *A killer whale spyhops, taking a brief glance at the surrounding terrain of San Juan Island near Washington state.*

MARINE MAMMALS

MARINE MAMMALS

Above: *The spinner dolphins get their name from their incredible acrobatic feats.* Below: *This young spotted dolphin will develop the pattern its named for when it matures.* Opposite page: *Like most dolphins, spotted dolphins aggregate in pods.*

Intelligence in Dolphins

People have long claimed that dolphins are very smart. A highly social lifestyle, complex playing behavior, and curiosity toward humans give strong support to this idea. Stories of dolphins saving the lives of distressed swimmers prompt some to wonder just how intelligent they might be.

While science has not yet studied all aspects of dolphin behavior, they are clearly quite intelligent animals. Their ratio of brain size to body weight is very high. In captivity, they quickly learn complex tasks, develop close attachments with their care givers, and show remarkable problem-solving skills. Their behavior toward each other in the wild also supports this idea, particularly the sophisticated communication that goes on among members of the same species. Dolphins use a complex series of sound, swimming patterns, and body language to relay messages to each other.

MARINE MAMMALS

MARINE MAMMALS

MARINE MAMMALS

Echolocation in Marine Mammals

Echolocation is a form of sonar used by a number of marine mammals, primarily the toothed whales. In echolocating, marine mammals learn about their surroundings by emitting sound waves and then receiving their reflected echoes. Echolocation is one of the most advanced sensory skills in nature.

When echolocating, toothed whales emit a series of clicks that they focus in tight beams. The sound waves strike various objects, and many of the waves bounce back to the animal that created them. In dolphins, the sound waves originate in the nasal passages. Reflected sounds are received by a fat organ in the lower jaw, called the mandibular fat body.

Echolocation may aid in navigation, hunting, and defense. Experiments have shown that toothed whales can tell the difference between objects whose texture, density, and internal structure differ only slightly. They can also sense the direction and speed that an object is moving.

continued from page 216

Spinner dolphins get their name from the incredible aerial displays they sometimes perform. They are fast and agile swimmers, and they often leap from the water, spinning and flipping before they crash back into the sea.

Several pods of spotted dolphins that inhabit the waters of the Little Bahama Banks have shown a strong curiosity for divers and swimmers. With no food rewards, these animals have repeatedly sought contact with humans. For the first few years, the contacts were short and involved only the large males. Male dolphins are typically more bold. Eventually, though, females and even juveniles came to approach humans in encounters that have lasted for hours.

Opposite page: *A pod of wild spotted dolphins in the Bahamas have become very friendly toward divers.* Below: *The major difference between dolphins and porpoises is that the porpoises have much shorter, blunter snouts. They share most other aspects of physiology and behavior.*

LIFE IN CORAL REEF COMMUNITIES

Very few natural settings are as stunning as the oceans' coral reefs. The massive coral structures, the dazzling colors, the subtle beauty of the reef creatures, and the diverse life within the coral kingdom are truly overwhelming. Excepting tropical rain forests, coral reefs are the most diverse habitats on earth, sheltering a vast number of marine species.

On entering a reef, the huge, buttresslike coral heads first grab a diver's attention. These bulwarks tower impressively in the turquoise water as they rise off the ocean floor. In many tropical seas, the normal visibility exceeds 100 feet, and on the best days, 300-foot visibility is not uncommon. In the clear waters of tropical reefs, snorkelers and divers will often sight coral head after coral head for as far as they can see or swim.

Opposite page: *A pair of butterfly fish amble over hard corals in the Red Sea.* Above, left: *A small school of snapper cruises under a rocky outcrop of a Caribbean reef.* Above: *Countless small fish create a moving veil over and around reef corals. The reef community is composed of a varied and spectacular range of vertebrates and invertebrates.*

In moments it becomes obvious that a coral reef is truly a community of organisms. Almost every nook and cranny houses one or more animals. Dense schools of brightly colored grunts, snappers, surgeon-

LIFE IN CORAL REEF COMMUNITIES

This page, top: *This large species demonstrates the startling color patterns tropical fish are famous for.* Above: *A secluded rocky outcrop houses a hidden cache of sessile, or immobile, bottom dwellers.* Opposite page: *Beige fire corals edged in white, red sponges, and soft corals grow side by side, attached to the bottom, while blue chromis and fairy basslets swim among them.*

fish, tangs, squirrelfish, and soldierfish swarm over Caribbean reefs, while schooling jacks cruise the upper reaches of the deep walls. Many schools contain hundreds, even thousands, of individuals. Yet each fish moves in perfect unison with all the rest as they travel the coral maze.

Parrotfish, filefish, puffers, and sergeant majors prefer smaller groups. Other species ranging from tiny blennies to the 20-foot great hammerhead sharks usually go it alone. Stealthy predators like groupers, 600-pound jewfish, and other seabass also live a solitary existence; they have only one mouth to feed if they hunt alone. Barracuda, various reef sharks, and moray eels are other important predators in coral reef communities.

For many new snorkelers and divers, it seems that the reef's only creatures are the hard corals and the dazzling fish. The imposing coral heads dominate the vista, and the ornate, vivid designs of many fish are so striking that it is difficult to see anything else. Soon, however, reef explorers will find the intricate beauty of the invertebrate world. Brilliantly colored sponges, sea fans, sea whips, and soft corals provide unforgettable visual treats. A close inspection reveals magnificently colored nudibranchs, crabs, brittle stars, worms, sea urchins, snails, scallops, oysters, anemones, and many other prominent members of the community. Less obvious, perhaps, but equally important are reclusive animals such as shrimps, lobsters, octopi, and squid.

LIFE IN CORAL REEF COMMUNITIES

LIFE IN CORAL REEF COMMUNITIES

Not all fish are as colorful or obvious as those first encountered. Potent scorpionfish and stonefish hide in the reef area, while trumpetfish find shelter in soft coral trees, hoping to go unnoticed. Another camouflage artist, the peacock flounder, hides in the sandy areas between the coral heads.

Though at first glance the sand patches and rubble zones in coral reef systems often seem barren, these areas are an important habitat for many species. Goatfish and hogfish excavate the seafloor, digging and blowing sand away as they hunt small mollusks and crustaceans. Sand tilefish, gobies and blennies, garden eels, razorfish, and some seabass also reside in the sand and rubble.

Stingrays belong to the sand community as well. Their close cousins, the spotted eagle rays, commonly patrol the vertical walls near the reef's seaward edge. Here, where the open ocean collides with the reef, is a likely place to encounter dolphins, sharks, and other creatures that inhabit deeper waters.

Sand flats next to coral reefs have their own community of organisms. Opposite page: *The undulating wings of a southern stingray send it gliding over the sand, as a jack keeps pace overhead.* Below: *A group of garden eels, rising from their burrows to face into the current to feed, make an eerie underwater sight.*

LIFE IN CORAL REEF COMMUNITIES

LIFE IN CORAL REEF COMMUNITIES

THE REEF STRUCTURE

Coral reef communities occur primarily in two sectors of the world: the tropical waters of the western Atlantic Ocean, which includes the Caribbean Sea, and the tropical Indo-Pacific, a region that stretches from the east coast of Africa and the Red Sea to Australia and the central Pacific Ocean. Other isolated coral reefs have taken hold near Hawaii, in Mexico's Sea of Cortez, and along Central America's west coast. All major coral reef systems are on the eastern side of continents. While there are some on the western sides, they pale by comparison in size.

Unlike any other habitat on earth, coral reefs form over countless millennia through the efforts of millions and millions of tiny animals. Corals are dynamic, living organisms. Every hour, every day, some coral polyps perish and others spring to life. In a slow but steady process, coral reefs grow and alter their shape ever so slightly. Living corals thrive on the outer edges of coral heads, but the reef's foundation is the limestone skeletal remains of past coral generations.

As living animals, corals require certain conditions to survive. Water temperature and water clarity are critical. True reef-building corals, or hard corals, can flourish only in shallow, clear waters where the temperature generally exceeds 70°F. Tropical seas provide the perfect environment for them.

Coral reef systems are far more than coral heads growing close to one another. The

Opposite page: *Hard corals that create large branching or mounding colonies are responsible for building the structure of the reef.* This page, top: *The skeletons of cup corals are too small to contribute substantially to the reef structure, but they are a part of the reef community.* Above: *Each polyp in this pair of solitary corals produces its own skeleton.*

LIFE IN CORAL REEF COMMUNITIES

230

LIFE IN CORAL REEF COMMUNITIES

Oh! Those Spectacular Walls

In many coral reef systems, the seafloor slopes softly toward the depths, slowly deepening as the bottom moves away from shore. In other areas, vast walls plummet sharply into the depths at the reef's seaward edge. These walls, or drop-offs, can be so sheer that they are almost vertical.

The drop often begins a hundred feet or more below the surface. Just off the Caribbean Island of Little Cayman, however, the top lip of the famed Bloody Bay Wall is only 20 feet from the water's surface. In the South Pacific and Red Sea, many drop-offs begin in very shallow water. Divers and swimmers can be standing in waist-deep water, take one step seaward, and hover thousands of feet over the seafloor.

The walls create some of the world's most spectacular seascapes. They support a wide range of colorful sea fans, sea whips, corals, sponges, and other invertebrates that harvest the food-laden currents sweeping the walls. Dense schools of colorful tropical fish swarm along the walls as they search for food. Perhaps the most dramatic feature of diving along a drop-off is to turn, back to the wall, and face the open sea. Any moment could bring a pod of dolphins, a shark in its constant search for food, or an enormous school of hungry tuna or jacks. Hovering over the dark ocean depths, backed by the lush reef, and facing the endless open sea, even experienced divers are left breathless.

reefs are complex structures. An entire reef system is a great labyrinth of caves, crevices, tunnels, sheer walls, and protected lagoons, all created by the corals' individual growth patterns.

Reef-building hard corals occur in various shapes and sizes. Most reef communities have quite a mixture of species. Their common names—elkhorn, staghorn, star, flower, brain, mushroom, pillar—tell their general shape. In many instances, the different species grow quite near one another. However, as corals grow, they compete for food, space, and sunlight. Active battles occur between the competing species. Some corals extend long filaments that kill and digest the tissue of intruding neighbors.

Opposite page: *A scuba diver in the southern Pacific Ocean encounters one of the myriad breathtaking sights on a reef's seaward wall.* Above: *This whitetip reef shark is one of the dominant predators in its community near Hawaii.*

LIFE IN CORAL REEF COMMUNITIES

Threats to a Reef

Corals have many living enemies. Certain fish and sea stars like the crown-of-thorns feed directly on coral. Other animals, mostly worms, sponges, and snails, burrow into the reef and weaken its structure. The most serious dangers to a reef, though, come from other sources.

As a living reef develops, it must contend with the merciless open sea. The sea is normally calm and gentle. At times, though, the frenzied power of a tropical storm sweeps the ocean, unleashing brutal force against the reef.

Major storms can devastate a reef, but even minor squalls miles away can create swells that uproot coral heads. Not all the broken coral perishes, but certainly much damage is done. Algae can move in quickly and prevent larval coral from starting new colonies. It can take years for a reef to recover from a direct hit by a major storm.

Today, man-made threats of silt and oil spills pose serious problems to healthy coral reef communities. Silt is often a by-product of construction projects and the deforestation of the land. Both silt and oil can quickly suffocate corals and wipe out entire reef systems.

When two competing species fight for survival, the outcome is sometimes predictable. Specific varieties tend to dominate certain other species. The corals that lose direct fights survive by other means. They often grow and reproduce faster than their more potent neighbors. The faster-growing species sometimes have better access to food-carrying currents or they block the sunlight from the slow growers and thus gain the advantage.

The end result of these battles is a reef system that harbors many different corals and many different habitats, providing food, shelter, and space for many different animals. Each coral species occupies its own niche in the reef system. Over time, their growth patterns form the reef's different zones—the seaward wall, the fore reef, the reef crest, the back reef, the rubble zone, the sand patches. Different species of marine wildlife settle in the zones that provide the living conditions they need.

This page: *A Spanish hogfish heads toward the floor of a reef to search for potential prey buried in the sand.* Opposite page: *A huge sea bass lurks just above the bottom. Sea bass of this size tend to be more sedentary than smaller ones.*

LIFE IN CORAL REEF COMMUNITIES

233

LIFE IN CORAL REEF COMMUNITIES

LIFE IN CORAL REEF COMMUNITIES

DIVERSITY IN THE REALM OF THE CORAL REEF

The coral reefs in the Indo-Pacific differ greatly in their diversity from the reefs of the Atlantic and the Caribbean. The tropical Indo-Pacific waters house more than 700 coral species, while only 60 or so species live in the Atlantic and Caribbean. More than 500 fish species make their homes in the Atlantic reefs. The number is large, but it would be only a meager sampling of the Indo-Pacific's fish. The tropical reefs of New Guinea alone have more than 1,500 species of fish. Another 1,500 species inhabit the Phillipines' reef communities. Still another 1,500 prowl Australia's Great Barrier Reef and the coral reef communities between southeast Asia and Australia.

The Indo-Pacific reefs' greater range of life comes in part from their great age. It has been much longer since any major geologic events restructured the Indo-Pacific. Thousands of years ago during the ice ages, temperature fluctuations had a greater impact on the Atlantic basin than on the waters of the west. Many living Caribbean reefs are less than 20,000 years old, while their Indo-Pacific counterparts are much, much older. The Indo-Pacific region is also a great deal larger than the Caribbean. The additional space allows a wider variety of species to evolve.

Coral reefs are certainly prolific, but the marine regions that surround them are relatively barren. Surprisingly, even the forbidding polar seas contain more life than

Reef fish are all different, each one adapted to its own niche. Opposite page: *Many small fish have no defenses of their own. They often school to confuse predators.* This page, top: *A butterfly fish pokes its long, tapered snout down into a coral to pull out a small crustacean.* Above: *Triggerfish have a sturdy dorsal spine that they can extend and lock in place, wedging themselves deep within a reef crevice when frightened.*

LIFE IN CORAL REEF COMMUNITIES

LIFE IN CORAL REEF COMMUNITIES

Australia's Great Barrier Reef

Located off Australia's northeast coast, the Great Barrier Reef is over 1,250 miles long and 200 miles wide at points. It covers an area of approximately 80,000 square miles. The next largest barrier reef in the world, the Caribbean's Turneffe Reef, is only about 125 miles long. The individual corals that compose the reef are about the size of a quarter or smaller. Nature needed at least a million years to create this massive structure.

While it's size and age are remarkable, the real beauty of the Great Barrier Reef is the life that thrives there. More than 350 species of coral and 1,500 species of fish live in the reef community. The richness of animal life ranges from single-celled plants and animals to large sharks. It is truly the most spectacular of all reef communities.

Opposite page: *Green tube sponges, feathery soft coral, and blue chromis make up just one of an infinite combination of seascapes.* This page, top: *Perhaps just a few feet away, blue tangs and grunts swim over the hard corals.* Above: *A crown-of-thorns sea star is attacked by one of its few predators, a triton trumpet. Coral-eating crown-of-thorns have devastated many reefs in the Pacific Ocean.*

the open waters of the tropics. Coral reefs are unmatched in their variety of life, but in sheer numbers, tropical regions as a whole have sparse populations.

Water clarity is an obvious difference between tropical and temperate shallow waters. Visibility can extend hundreds of feet in the tropics, while divers in temperate and polar seas often can see only thirty or forty feet. Because of the limited view, life easily goes unseen in the colder waters. Even experienced divers don't always know that these regions house a far greater number of animals.

Plankton, microscopic plants and animals, cause part of the murkiness in cooler marine

LIFE IN CORAL REEF COMMUNITIES

This page, top: *This bed of corals in the Red Sea contains a variety of species, each contributing its unique growth pattern to the reef structure.* Above: *An army of cleaner shrimp swarm over a pair of California moray eels. The shrimp will clean the eels of parasites and dead tissue.* Opposite page: *Some reef fish go through dramatic color changes when they mature. This juvenile french angelfish will lose its bold yellow stripes.*

regions. They are also responsible for the profusion of life there. In unimaginable number and mass, these tiny organisms cloud the waters and form the basis of a complex food chain that supports huge populations.

Tropical open oceans do not support anywhere near as many microscopic plants and animals. Lacking this solid base for food chains, they are actually vast, watery deserts. A coral reef community is somewhat like an oasis in this desert. It provides a rich habitat filled with dense populations and a wide range of species.

THE DYNAMICS OF REEF LIFE

The reef residents spend their lives interacting. Perhaps the most important reef relationship exists between the algae called zooxanthellae and the corals in which the algae lives. Through photosynthesis, the zooxanthellae produce nutrients. The corals give the algae nutrients through waste products, and the algae provide the corals with the raw materials for building their skeletons. The corals survive and create the backdrop for reef life, and the algae survive and form the basis of the food chain that supports reef populations.

As corals grow, the reef becomes vulnerable to animals that bore into dead and older corals. Various sponges, snails, and worms burrow into the reef and take up residence, weakening the reef and causing coral heads to collapse. The collapsed corals provide a surface where new corals can restart the reef building process.

LIFE IN CORAL REEF COMMUNITIES

239

LIFE IN CORAL REEF COMMUNITIES

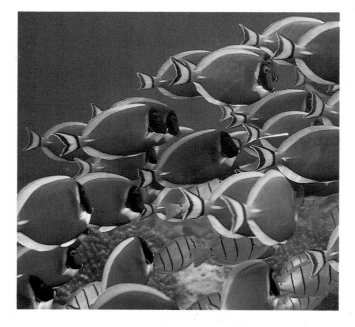

Above: *A mixed group of fish travel over a reef in the Maldives.*
Below: *An eagle ray lazily cruises through the Red Sea.*
Opposite page: *A silversided damselfish guards its territory, a lawn of algae.*

Some fish stake out and defend small territories in the reef. Damselfish cultivate algae within their territories. They bite the living corals, creating weak spots that become overgrown by algae, and claim the area containing the growths. The damselfish ends up with its home, but the corals must suffer. Parrotfish also cause damage to the reefs. As parrotfish feed, they actually scrape the surface of the coral with their beaklike teeth. This weakens the coral, making it vulnerable to boring sponges, snails, worms, and other animals.

However, parrotfish and other species that consume coral or the corals' algae help the reef community in other ways. These fish produce much of the sand found near coral reefs as waste product. Creatures that live or hunt in the sand have their habitat partly created by the corals' enemies.

continued on page 244

LIFE IN CORAL REEF COMMUNITIES

LIFE IN CORAL REEF COMMUNITIES

242

LIFE IN CORAL REEF COMMUNITIES

The Ocean Reclaims

Competition for living space is intense among marine organisms. Losing the battle for space can mean losing the battle for life. When new living space opens up, marine species are quick to take advantage. Shipwrecks and other artificial structures can be ideal places for marine communities to take hold.

Shrimps, snails, crabs, barnacles, hydroids, and worms are usually among the first animals to stake a claim. Juvenile fish and larval fish that settle from the plankton see new structures as places where they don't have to compete with established adults. As soon as algae patches start to grow, often in less than a month, foragers like parrotfish and surgeonfish move in. They attract larger predators, first jacks, tarpon, and barracuda that roam over wide ranges, and later others.

Corals are actually among the last animals to appear in artificial reefs. They prefer to settle on limestone-based solid surfaces. It usually takes a year or so before enough limestone accumulates to suit even a few species.

In some cases, algae can cover artificial reefs within 10 days, and after only two weeks, as many as 15 fish species might appear. A great many ships sunk near Micronesia's Truk Lagoon during World War II. Today, less than 50 years later, the mix of life around these wrecks is almost identical to the plants and animals of nearby natural reefs.

Opposite page: *Shipwrecks attract scuba divers as well as marine life.* This page, top: *Under water, every hard surface becomes encrusted with life. In Truk Lagoon, the devastation of war becomes a reef teeming with life. This Japanese ship, the* Fujikawa Maru, *was a casualty of World War II.* Above: *Many ships are sunk by accident and in war, while others are intentionally sunk as artificial reefs.*

LIFE IN CORAL REEF COMMUNITIES

continued from page 240

The coral reef scene is quite different during the day and during the night. Many fish, especially the more colorful ones, are far more active during the day hours. As the sun sets, fish that feed during the day seek the reef's protective cover. Angelfish and damselfish hide in crevices. Some parrotfish surround themselves with mucus that conceals their scent from predators like moray eels.

Species such as the squirrelfish and cardinal fish prefer to hunt in darkness. The whitetip reef shark and many other reef sharks are far more active at night. Invertebrates such as shrimp, hermit crabs, bristle worms, and squid openly roam the reef to hunt and forage in the evening.

Below: *A parrotfish rests within a blanket of mucus.* Opposite page: *A large spider crab, also called the Caribbean king crab, lurks in a Cozumel reef cave.*

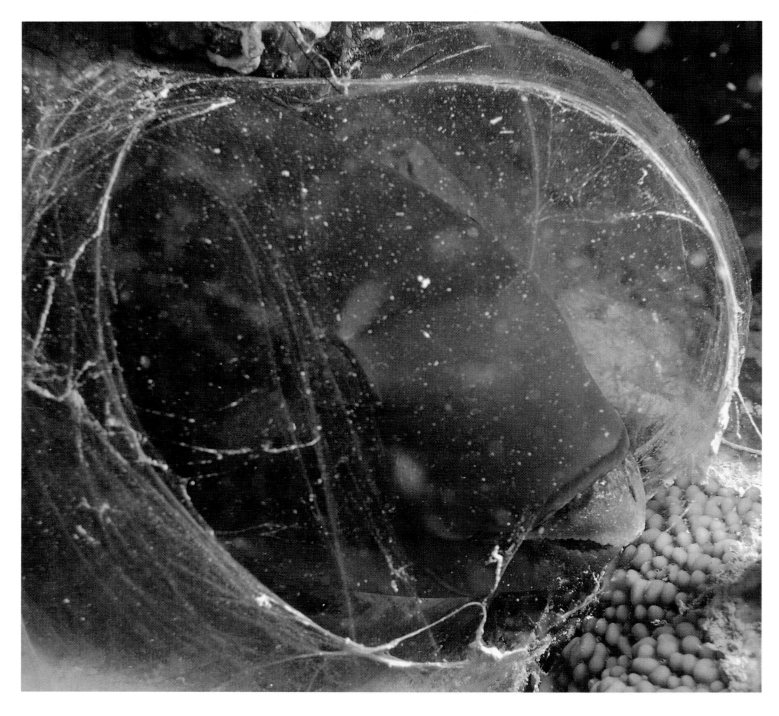

LIFE IN CORAL REEF COMMUNITIES

LIFE IN CORAL REEF COMMUNITIES

LIFE IN CORAL REEF COMMUNITIES

The twilight hours just before and just after sunrise and sunset are especially active times. These transition periods see one group of feeders retire to safety while another group begins to roam the reef. The reef inhabitants are perhaps more alert and tense than usual as the water fills with ever-hungry predators. Jacks, barracuda, and other predators actively hunt in these low-light conditions.

This daily cycle is a vital part of the complex coral reef community. The cycle enables reef dwellers to best use the available space. Except for areas claimed by territorial creatures, a reef's nooks and crannies house both nighttime and daytime occupants. Shortly after a nocturnal squirrelfish leaves a protective reef crevice, a day-feeding angelfish might slip into the same hole to rest safely for the night.

Opposite page: *A carnivorous ewa blenny, half in and half out of a hole in a Fiji sponge, is ready to snap up a meal or retreat to safety.* This page, top: *A group of squirrelfish choose this small cave to rest in during the day.* Above: *A wrasse blenny spends most of its time peering from its burrow but will sometimes venture out to swim with schooling bluehead wrasses.*

LIFE IN TEMPERATE SEAS

The temperate seas are vast expanses of ocean lying between the tropical and polar regions. The surface water temperatures there avoid the extremes of other marine habitats, ranging from the low 40s to the low 70s. Reef-building corals cannot survive the cool waters, so rocky areas and kelp forests serve as a primary haven for marine life.

A region's environmental conditions determine its spectrum of life. The rule applies to temperate seas as it does to all natural habitats. Water temperature, currents, weather patterns, and geographical features are different in temperate waters than in all other ocean environments. The creatures that live there differ accordingly.

Conditions in the temperate zone are usually not as inviting for underwater explorers as they are in the tropics. The water is colder and often much rougher, and underwater visibility is usually poorer.

Because conditions are not as accommodating for humans, people often assume that temperate seas are not as prolific or interesting as the tropics. Nothing could

A multitude of life thrives in the cool waters of temperate seas. Opposite page: A brown rockfish rests on the sea bottom just off Monterey Bay, California. Above, left: Farther out to sea, the roof of a giant kelp forest is awash in the water's surface. Such a canopy is a clear sign that life abounds below. Above: Beneath a kelp canopy, a kelpfish swims among the huge plant fronds near Catalina Island.

LIFE IN TEMPERATE SEAS

This page, top: *Like living pillars in the sea, giant kelp fronds reach from the seafloor up to the surface and the life-giving sun. The space between the kelp teems with fish and other marine life.* Above: *Two club-shaped rhinophores mark the head end of a striped armina walking along a kelp plant. These nudibranchs, or sea slugs, more commonly burrow just below the surface of a sandy or muddy seafloor.* Opposite page: *At the base of a giant kelp forest in the Channel Islands, orange-gold garibaldis stand out in contrast to the green-brown kelp and the muted colors of many other forest inhabitants.*

be further from the truth. Marine life abounds in temperate seas. In many areas, rich, cool waters support dense populations of microscopic organisms, which in turn support an astonishing variety of marine creatures. The variety and richness of marine life in temperate oceans makes many tropical seas appear to be veritable deserts.

FORESTS WITHIN THE SEA

A feature attraction of some temperate oceans are their magnificent kelp forests. Kelp is a form of brown algae that most people only see once it has torn loose from the seafloor and washed ashore to decay. Growing in the ocean, though, a healthy kelp forest, or bed, rivals the beauty and wildlife of any natural setting.

On the best days, when the surface waters are calm and the sky is clear, a diver couldn't wish for more. Underwater, the rayed sunlight dances through the golden kelp beds, creating a cathedrallike effect as waves pass gently overhead. The towering plants stand out against a background of blue-green water. At a depth of 30 to 40 feet, divers notice the forest's rhythmic sway as each plant rocks gently with the passing swells. Along the bottom, strands of bright green eel grass and other marine vegetation flow back and forth with the surge.

Keen observers might glimpse a camouflaged kelpfish as it meanders through the blades of the kelp plants. Kelpfish are golden, knife-shaped fish; their color and outline look much like the blade of a kelp plant. If it were spring, divers might look
continued on page 254

LIFE IN TEMPERATE SEAS

A Closer Look at Giant Kelp Plants

Like other species of kelp, giant kelp requires a rocky surface so the plant can secure itself. Kelp has no true root system like terrestrial plants, but instead uses thin, sturdy structures called haptera to grip the bottom. The haptera look much like oversized spaghetti. The strands do not penetrate the seafloor. Instead they grip and adhere to rocky surfaces and form a holdfast that secures the plant.

Mature plants consist of a holdfast and many buoyant fronds. The fronds have a connective stemlike stipe, many blades, and a series of water-tight gas bladders. Together, the gas bladders buoy the fronds and prevent them from sinking to the seafloor. The plants then receive the sunlight that is so vital to all plants' survival.

Giant kelp has adapted well to its life in the sea, flourishing in conditions that would prove difficult for many other plants. In normal sea conditions, the kelp blades constantly sway in the currents and swells, facing one direction one moment and another direction the next. Both sides of kelp blades are able to produce nutrients with the aid of the sun, so the plants can sustain themselves despite their constant motion. In addition, the kelp blades are wrinkled, greatly increasing their surface area and the plant's ability to absorb sunlight. These adaptations play a significant role in making giant kelp the ocean's largest and fastest growing plant.

Kelp plants are highly flexible, which helps them survive one of their greatest natural threats. The violent surge and swell that accompanies heavy storms can destroy a kelp forest. The flexible quality lets the stipes move with the heavy water motion instead of fighting against it, so that healthy plants are not torn from the bottom.

This page: *Gas-filled bladders at the base of each giant kelp blade keep the plants from sinking. Suspended thus in midwater, the blades can carry out their vital task of gathering sunlight. Giant kelp, which can grow more than 100 feet a year, is the largest marine plant known.* Opposite page: *Kelp forests attract California sea lions as well as fish and invertebrates. The sea lions see the forest as a rich source of food and also as a place of relative safety. The dense vegetation can impair the hunting skills of one of their primary predators, the great white shark.*

LIFE IN TEMPERATE SEAS

LIFE IN TEMPERATE SEAS

continued from page 250

down at the rocky bottom and see a 14-inch bright orange male garibaldi dash from a crevice to defend its nest. And if they glance toward the surface canopy, the divers might see a sea lion or harbor seal chase a shimmering school of silver-colored anchovies. As divers pause, look, and admire, they will see an astonishing sampling of life.

In rough and stormy conditions, though, few places appear more uninviting and ominous. As the sea churns, the towering fronds violently tug at the holdfasts that secure them to the bottom. Many kelp plants pull free from the bottom and entangle neighboring plants, which in turn tear loose. In such conditions, a diver would whip about underwater like a leaf in a gale.

These scenes are from a forest of giant kelp, one of hundreds of kelp species found in temperate seas. Giant kelp plants grow to 200 feet, but their base is usually set 80 feet or less below the surface. From the rocky bottom, giant kelp plants grow straight up toward the surface. Once the frond reaches the surface, the excess length spreads out and floats, forming a lush surface canopy. Giant kelp plants typically grow near one another, forming dense undersea forests up to 10 square miles.

Below: *This gnarled kelp holdfast is designed to grip the rocky bottom tenaciously during violent swells and storms.* Opposite page: *In the Pacific Northwest, a tealia anemone spreads its pink-tipped tentacles and awaits a chance encounter with something edible. Just behind the anemone are two red urchins, a brightly colored fish, a mass of sponges, and various forms of plant life.*

LIFE IN TEMPERATE SEAS

LIFE IN TEMPERATE SEAS

A healthy forest of giant kelp can be home to more than 800 species of marine animals. A single plant can support more than a million organisms. Many are microscopic, but the sheer numbers show just how prolific a healthy forest can be. A column of water containing giant kelp plants supports thousands of times more organisms than the water above a barren, sandy bottom.

Scientists recognize three habitats within a kelp forest. The holdfast habitat contains many microscopic organisms as well as crustaceans, brittle stars, sea stars, hydroids, bryozoans, snails, and sea cucumbers. These creatures use the holdfast habitat as a place of attachment, a place to hide, and a source of food. Above the holdfast lies the mid-water habitat where giant kelpfish and kelpbass hide in the flora. Clingfish and numerous invertebrates attach or cling to the mid-water portion of the plant. The third habitat is the surface canopy that fish, crabs, mollusks, and other invertebrates call home. The canopy is also a favorite place for sea lions, elephant seals, and harbor seals to frolic and feed.

Scuba divers who explore giant kelp beds regularly see creatures ranging from inch-long nudibranchs to 40-foot, 35-ton gray whales. Within that spectrum, divers find a myriad of marine life to enjoy. Kelp surfperch, opaleye, and mackerel live in large schools, while garibaldi and cabezon prefer a more solitary existence. Some species of rockfish hide among the kelp fronds. Other specimens hover near the rocky bottom below.

Opposite page: *A pair of kelp snails move over a fallen frond, probably feeding as they go.* This page, top: *Kelp crabs not only match kelp in color, but they munch on bits and pieces of the plant.* Above: *This female sheephead is a general feeder. Using its stout jaws and large teeth, it will consume most any medium-sized invertebrate it finds, even hard-shelled abalone and crab or a sharp-spined urchin.*

Above: *Sea otters abound in kelp forests where they can find abundant supplies of their favorite foods, including sea urchins. They often bring their prey to the surface to feed, lying on their backs amid the kelp canopy.* Opposite page: *Purple sea urchins can wreak havoc on kelp by eating through their holdfasts. No longer anchored, the plants drift free in the water, snagging and ripping up other plants.*

Dense schools of migratory barracuda and yellowtail visit the kelp beds, as do solitary species like white seabass and gigantic black seabass. Black seabass can exceed 7 feet and at that length will likely weigh more than 500 pounds. Majestic bat rays often cruise through the kelp beds. Bat rays are among the most graceful swimmers, and their powerful jaws enable them to crush clams, crabs, and other shellfish.

In southern California's giant kelp beds, sheephead search the reef for spiny sea urchins and other prey. Attaining a weight of 40 pounds or more, sheephead are large members of the wrasse family. During the first stages of their adult lives, all sheephead are females. As they mature, many will become males to meet the local population's needs. Juveniles are reddish orange with prominent black spots on their body and fins, and mature females are pinkish. The males have a squared, dark blue to blackish head and a red mid-section, while the rest of the body is bluish-black. All adults have a white chin.

California sea lions and harbor seals cavort in many kelp forests. These mammals feed on fish found in or near the forests, and in turn, they provide food for great white sharks, killer whales, and other top-end predators.

Further north, sea otters play a very important role in kelp forest communities. Because they lack the thick fat of other marine mammals, sea otters must eat copious quantities of food to help them stay warm. Feeding on crabs, urchins, lobsters, abalones, and more, immature otters can consume 35 percent of their body weight every day. The adults can devour up to 25 percent of their body weight or more each day. With their huge appetites, the otters are an important factor in the forest food chain.

LIFE IN TEMPERATE SEAS

LIFE IN TEMPERATE SEAS

Opposite page: *A relative of the sea horse and pipefish, this leafy sea dragon is quite comfortable in the kelp forests of Australia. The leafy, kelp-colored appendages that sprout all over its body create a perfect means of camouflage.*

Although the species of dominant kelp are quite different, the kelp forests of south Australia are remarkably similar to those in other regions. Relatives of the sea horse, leafy sea dragons and common sea dragons hide among the kelp fronds. They are slow swimmers, but their shape and coloration hides them in the low-lying kelp. Tommy Roughs fill a similar niche to that of anchovies in southern California.

The Delicate Balance

In nature, very few things are constant. Populations of predators and prey are always in a state of flux, for example. As food sources expand, the populations that use those sources often expand in response, and when food sources diminish, related populations will shrink. In nature's complex web, such events can have far-reaching effects on a habitat.

In a kelp forest, sea urchins normally feed on drift kelp, the dying and decaying blades dropped by healthy plants or plants dislodged by waves and storms. However, when the competition for food is stiff, sea urchins readily turn to the healthy kelp plants. They eat through the plants' holdfasts, and the kelp pulls free from the bottom, often entangling and damaging other plants. Urchins can be voracious eaters; if left unchecked they can devastate a kelp forest.

Sea otters prey heavily on the urchins, and in doing so are protectors of the kelp forests. As a part of the kelp forests' life cycle, kelp, urchin, and otter populations fluctuate within natural norms. Otter populations may wane, and the decreased pressure on the sea urchins allows their numbers to grow. In time, the abundance of urchins means a rich food source for the otters. The otter population and its need for food increase, and they once again check the urchins' population growth.

In the early 1900's, human intervention jeopardized many California kelp forests. For years, hunters and trappers intensely pursued sea otters for their pelts, and the otters neared extinction. With the otters gone, sea urchins rapidly expanded all along the California coast. The overpopulated urchins turned to healthy kelp plants as a food source, and thousands of square miles of kelp vanished from the sea. Along with the kelp went the habitats of countless marine creatures.

In the 1960's, replanting programs helped restore many kelp beds, and otters were reintroduced in some areas. Still, the issue raises controversy today. In addition to urchins, the otters consume lobster and abalone. The commercial fishing industry sees them as direct competitors and objects to efforts that increase the otter populations.

LIFE IN TEMPERATE SEAS

LIFE IN TEMPERATE SEAS

THE ROCKY AREAS OF TEMPERATE SEAS

The large structures in the temperate zones consist of rock. These rocky areas provide a surface for attachment and caves and crevices for protection. The rocks provide a backdrop for lobsters, crabs, snails, scallops, and other animals. They also serve as wave breaks, providing respite from the violent, churning open sea.

Competition in these communities is intense. Predatory sea stars hunt clams, scallops, and other bivalves, while some mollusks search for snails and nudibranchs. Many fish pursue urchins, scallops, sponges, crabs, lobsters, shrimps, and octopi. Some nudibranchs and many snails graze on the algae that grows on the rocks.

Often, a variety of brightly colored sea anemones cover the rocks. Some anemones live their lives as solitary creatures, while others live in groups. If potential prey, perhaps a small fish, touches the anemone's tentacles, the anemone quickly stings the fish, wraps it in its tentacles, and draws it into its mouth.

Other residents include sponges, barnacles, worms, and small, colorful fish such as blennies and rockfish, and more. In areas where currents prevail, barnacles, scallops, and sea whips flourish. These creatures affix themselves to the rocks and trap prey from the passing currents. Where currents are strong, a diver might see rows and rows of red, yellow, orange, and golden sea whips. Yet only a few yards away, in more protected waters, there might not be a single one.

Opposite page: *Goose barnacles thrive in the rocky areas of the Pacific Northwest. They congregate on almost any hard surface; occasionally they wash ashore still attached to bottles, wood, or fishing floats.* This page, top: *Amid a tangle of pink arms, the central disks of several brittle stars stand out.* Above: *Rocky reefs perpetually serve as battlegrounds for life-and-death struggles, as evidenced by this octopus consuming one of its favorite foods, a snail.*

LIFE IN TEMPERATE SEAS

Above: *Juvenile garibaldis are the same bright orange as their parents, but they also have iridescent blue spots and trim. These juveniles may be one of the most stunning kelp forest fish.* Below: *Pacific barracuda inhabit both surface and deep water. Whole schools will relentlessly hound small baitfish, like sardines and anchovies.* Opposite page: *A voracious anemone waits for a fish to swim into its tentacles.*

The cracks and fissures of the rocks provide ideal living quarters for lobsters, octopi, shrimps, crabs, eels, and numerous other fish. Primarily nocturnal creatures, lobsters prefer to forage for food at night. During the day they hide in the rocks' recesses, often gathering by the dozens or even hundreds. The octopi usually live as solitary creatures in well-protected dens, and they too prefer to seek their prey at night. Octopi feed heavily on crabs, lobsters, and other crustaceans as well as other mollusks and some small fish.

The wolf eel is a fish that suffers from its intimidating appearance. At first glance the wolf eel's long, tube-shaped body appears very similar to the moray eel. Despite their common name and their shape, they are not true eels. They are actually close relatives of the blenny family.

LIFE IN TEMPERATE SEAS

LIFE IN TEMPERATE SEAS

Inhabiting the rocky bottoms, wolf eels have a mouth full of sharp canine teeth, but their looks belie their docile nature. The largest wolf eel ever documented measured just more than 6 feet long. Wolf eels hunt sea urchins, crabs, fish, and some mollusks.

Several species of bottom-dwelling sharks commonly inhabit the rocky areas. Surprisingly, most temperate water reef sharks, like the horn shark, swell shark, leopard shark, and Port Jackson shark, are very tame creatures. They don't resemble the typical image of teeth and power people usually attribute to sharks. To many people, horn sharks look more like catfish than sharks, but they are true sharks. Horn sharks, Port Jackson sharks, and swell sharks prey on various crustaceans, mollusks, echinoderms, and fish.

In the shallow regions of temperate seas, great white sharks play an important role as top-level predators. Due to their large size, many people believe that great white sharks roam the open seas. Actually, scientists consider them to be shallow-water sharks that prefer areas over continental shelves. As juveniles, great white sharks prey upon rays, flatfish, and other small marine animals. As they mature, their diet shifts to marine mammals such as seals and sea lions.

Opposite page: *Countless tentacles make this metridium anemone look like a fuzzy dust mop. Despite the delicate color and soft appearance, it is just as deadly as any other anemone.* This page, top: *The craggy face of a wolf eel appears grotesque and stonelike, but these creatures are actually quite docile.* Above: *Even this gaudy, eye-catching pink nudibranch can find a backdrop in the Pacific Northwest that makes it a bit less obtrusive.*

LIFE IN TEMPERATE SEAS

LIFE IN THE SAND

In many temperate regions, expanses of sand lie between the rocky areas. Some sandy patches extend hundreds of square miles, and others are only a few square yards. Regardless of size, almost all these areas teem with life.

In many respects the oceans' sandy plains bear a strong resemblance to the deserts found on land. During the day, the sand habitat can appear barren and lifeless. Lacking the fissures, caves, and crevices of a rocky habitat, the sandy plains offer fewer natural hiding places. Many creatures that live here feel safer in darkness and are far more active during the night.

Regardless of when they are active, the creatures of the sand must be able to defend themselves from their potential predators and to catch their prey. Like all other creatures, they live in a dangerous world, and like all other creatures, they have developed unique ways to counter the threats they face.

Many sand residents are adept burrowers and diggers that use the seafloor as cover. Small fish called cusk eels burrow into the sand during the day and emerge at night to find food. Clams, crabs, and shrimps not only dig well but can quickly rebury themselves if uncovered. Even the large predators—stingrays, electric rays, guitarfish, and angel sharks—use the sand for cover, burying themselves and exposing only their eyes.

Opposite page: *Hydroids live wherever they can find a suitable surface to which they can attach.* This page, top: *A copper rockfish hangs motionless for a moment over the sand.* Above: *A speckled sanddab blends expertly with its surroundings. Flatfish can hide by matching their surroundings or by burying themselves in the sand.*

Other animals create their own hiding places. Some worms, for example, build tubes to live in. The tubes secure them to the seafloor where the worms can take food from the passing currents, and they provide protection.

Many sand residents have colors and patterns that match their sandy surroundings, and some can alter their coloration to hide themselves. Halibut, turbot, flounder, sanddabs, and other flatfish alter the shape of special skin cells to control their color. They can change their splotchy skin pattern to better match their surroundings.

While they are not born this way, adult flatfish have both eyes on one side of the head. In their larval stage, they inhabit the open sea and have one eye on either side of the head. As the fish ages, it moves toward shore and settles in a sandy region. In a short time, the fish produces enzymes that allow one eye to migrate to the other side. The fish can then almost completely bury itself in the sand and still watch carefully for predators and prey.

Opposite page: *In the sand communities of temperate seas, some sea anemones anchor themselves by burrowing into the soft seafloor.* Below: *A small red crab, normally found in the open ocean, is besieged by a group of kelp scallops in a sand flat.*

LIFE IN POLAR REGIONS

Few places on earth are more forbidding than the polar regions. On the warm days, the temperatures approach freezing. The landscape seems carved of ice and snow. Some of the fiercest storms on earth ravage this land, with howling, frozen gales blowing in all directions at once. Bleak, inhospitable, and frigid, the Arctic and Antarctic circles seem designed to keep life from taking hold.

Despite the conditions, polar seas are some of the most life-filled ocean regions on earth. Much of the life is plankton, countless millions of tons of microscopic plants and animals. But these organisms support a network of higher animals that have conquered their harsh surroundings.

Crabs, lobsters, sea stars, barnacles, squid, and countless other invertebrates populate the polar seas. Huge schools of fish roam the Arctic and Antarctic waters, feeding and being fed upon. Mammals, too, are key members of the polar habitats. Seals, sea lions, walruses, polar bears, and whales dominate the region.

Near the ends of the earth, life is plentiful in the face of the bitter polar climate. Opposite page: King penguins mill around on South Georgia Island. Above, left: *A group of walruses venture into the icy waters off Round Island.* Above: Among the Arctic inhabitants is this colorful glass jellyfish.

LIFE IN POLAR REGIONS

The harsh conditions make polar life difficult to study. Humans have only glimpsed the beauty of this frozen world. The animals that live here, though, have faced a merciless enemy and learned to persevere.

THE POLAR ENVIRONMENT

Although the Arctic and Antarctic lie as far from one another as our planet allows, these regions are similar. They are both lands of ice, snow, and bitter cold. Both poles experience a cycle of extremely long days and then nights. The sun shines for six continuous calendar months, and when it sets, six months of darkness pass until the next sunrise. Of course, they have opposite day/night cycles. When summer sunlight shines on the Arctic waters, the South Pole experiences six months of darkness during polar winter.

Above: *A pair of dungeness crabs mate in the ocean southeast of Alaska.* Below: *Young salmon migrate from rivers to the open ocean where they will grow to adults.* Opposite page: *A polar bear playfully rolls across the Arctic ice, perhaps to cool off.*

LIFE IN POLAR REGIONS

LIFE IN POLAR REGIONS

LIFE IN POLAR REGIONS

The poles also have some important differences. The Arctic Ocean has a permanent cap of ice. Arctic waters are bounded by major land masses—Canada, Alaska, Russia, Greenland, and northern Europe—but the region lacks large sections of land. The Antarctic, on the other hand, is a continent, land topped by ice and surrounded by water.

While both regions are severely cold, Antarctica's climate reaches greater extremes. Temperatures commonly plunge more than 130°F below freezing. Antarctica's mountainous terrain holds over 90 percent of the world's ice, and incredibly violent storms ravage the land. Land conditions create an almost sterile environment, so the biological cycle revolves around the marine kingdom. Life clings to the edges of the continent and a few other isolated areas where slightly milder conditions allow a handful of plants and animals to survive.

In the Arctic, life expands and contracts with the region's ever-changing boundaries. The change of seasons either reduces or increases the amount of Arctic ice. Temporary chunks of ice called ice floes provide living quarters for many Arctic species. The floes can be a few square yards or several square miles. During winter, the ice floes reach the southern extremes of the Arctic Ocean, but as summer warms the water, the ice floes recede to the north.

Opposite page: *A penguin diligently tends to its young on a rocky outcrop.* Below: *A mother weddell seal has hauled out onto the ice next to her month-old pup on an Antarctic island.*

LIFE IN POLAR REGIONS

LIFE IN POLAR REGIONS

THE SEASONAL CYCLE OF LIFE

In both Arctic and Antarctic waters, the changing seasons govern the cycle of life. Sunlight is a crucial ingredient in this biological cycle, and the amount of sunlight cast upon polar seas varies sharply with the seasons.

During the summer, sunlight is abundant. Prevailing ocean currents and storms create deep-water upwellings, strong upward currents that bring minerals and other nutrients from the depths. Phytoplankton take these resources and grow. Summer in the polar regions sees these tiny marine plants reproduce in immeasurable volume.

Along with this plantlife comes the zooplankton, microscopic animals that feed on the tiny plants. Zooplankton, too, are quick to respond to advantageous living conditions. They multiply rapidly with the abundance of food, and the sea teems with a limitless food source for its larger creatures.

Animals ranging from bottom-dwelling sea stars, crayfish, and crabs to fish, seals, birds, and whales flourish in these conditions. Drawn by food, many migrate toward the poles during the summers and then travel toward the equator during the harsh winter months. The summer's bounty brings life to the polar regions, and the sterile winter drives much of it away.

Opposite page: *Penguins aren't the only birds that make Antarctica their home. These courting imperial shags will breed on an island off the Antarctic Peninsula. Imperial shags often form dense groups of thousands out at sea.* This page, top: *This icefish, a little over a foot long, rests on the seafloor in about 80 feet of water. Many fish that live in the south polar region have a sort of antifreeze in their blood that lets them survive the frigid waters.* Above: *Some invertebrates such as this brittle star have also adapted to the Antarctic climate.*

LIFE IN POLAR REGIONS

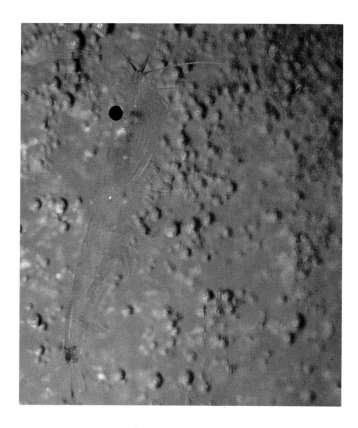

Above: *Krill, small crustaceans less than two inches long, make up the base of a productive Antarctic food web.* Below: *Penguins and elephant seals share a chilly stretch of land during the southern hemisphere summer. Both species are here to breed.* Opposite page: *The rockhopper penguin is easily identified by the spiked tufts and the splash of color that adorn its head.*

What Is Krill?

Strictly speaking, krill is a particular species of plankton, *Euphausia superba,* found only in the Antarctic. This tiny crustacean occurs in huge numbers and is one of the world's most important food sources. Huge shoals of krill feed countless whales, seals, fish, and birds that frequent the southern polar regions.

In common use, krill refers to a wide variety of tiny, shrimplike crustaceans. These creatures are similar to *Euphausia superba.* Found throughout the world, they too serve as food for many marine animals.

Many scientists feel that *Euphausia superba* may someday be a significant natural resource for humans. Krill is an excellent source of protein, and estimates say we could annually harvest 100 million tons of it—a good deal more than the world's current fish catch— without seriously damaging the krill population. In its natural form, krill would make an unappealing snack. Still, commercial processing could produce a krill-based protein concentrate to be mixed with common foods such as bread.

LIFE IN POLAR REGIONS

LIFE IN POLAR REGIONS

ANIMALS OF THE POLAR SEAS

Microscopic plants and animals provide the foundation of polar food chains. Still, many large and dramatic species play vital roles as well.

At least a dozen species of whales act in the ecology of Arctic and sub-Arctic waters. Filter-feeding blue whales, sei whales, minke whales, humpback whales, and bowhead whales consume enormous quantities of plankton in the water surrounding the North Pole. Blue whales are the largest creatures on earth. Even the smaller filter-feeding whales reach huge sizes. The minke whales, for example, commonly reach 30 feet and 6,000 pounds.

Toothed whales such as killer whales, beluga whales, and narwhals also inhabit Arctic waters. They feed on fish, squid, octopi, birds, and other mammals. Some of these species migrate to temperate and tropical zones regularly; others are rarely found outside Arctic waters.

Beluga whales inhabit shallow waters. They often swim in estuaries and bays, and during summer months they may travel more than 100 miles up some rivers. At birth, their bodies are dark brown with small bluish-gray spots. Within a short time, the spots fade, and belugas take on a stunning ivory color. As adults they are pure white.

Opposite page: *A pod of shy narwhals can only be seen in the Arctic seas. Each male wields a single spiral or twisted tusk.* This page, top: *A male killer whale cruises near shore, perhaps looking for a meal of fish or marine mammals.* Above: *A grey whale glides through the shallow coastal waters of British Columbia. Many whales come to these cool northern waters to feed during the summer months and return to warm waters to calve and mate during the winter.*

LIFE IN POLAR REGIONS

Known as the unicorn of the Arctic, narwhals are famous for their long tusks. The tusk, which is actually a modified tooth, can easily be six to seven feet long, almost half the length of the whale's body. Only the males possess tusks, and they probably use them to fight off predators and compete with other males.

Whales also play important roles in the Antarctic. Many species have separate populations that live in the northern or southern hemispheres. The northern and southern groups never mix because their yearly cycles are different. Even though the species are the same, subtle differences exist between the various populations, perhaps due to their slightly different courses of evolution.

Many whales and dolphins are also unique to the Antarctic and its surrounding waters. Among these are the filter-feeding southern right whales and pygmy right whales and hourglass dolphins, Commerson's dolphins, and spectacled porpoises.

Polar bears serve as a link between land and sea in the Arctic realm. On land, they reign supreme, the undisputed masters of the northern ice lands. Adult males can be 10 feet tall and 1,200 pounds. Polar bears are also excellent swimmers and spend considerable time hunting in and around the frigid waters.

Polar bears do not have permanent dens, but instead are tireless travelers. They live in solitude, except when mating or raising young. Their diet includes small land mammals, carrion, and eggs, but their favorite prey are ringed seals. Able to stalk prey over great distances, polar bears are unusually patient hunters. In winter, they wait by holes in the ice that ringed seals use to get air. When a seal surfaces, the bear attacks and quickly breaks its back. Their color enables them to blend in with their surroundings despite their massive size.

continued on page 289

Opposite page and above: *A polar bear, the largest terrestrial predator in the world, appears majestic whether resting on a snowbank or sitting up on its haunches.*

LIFE IN POLAR REGIONS

Serious Threats to Polar Wildlife

Polar wildlife is hardy if it is anything. Creatures that can thrive in the harsh Arctic and Antarctic climates are certainly resilient, resourceful, and enduring. Still, their lives are delicately balanced. Their bodies and habits have been honed to fit a specific environment. Like all creatures, changes in their habitat could bring extinction. Today, scientists know that polar ecology has felt humanity's presence even though we leave these areas relatively untouched.

Recent years have seen manufactured chemicals affect the polar atmosphere. Chemicals used in industry and everyday living have created holes in the ozone layers over the poles. These holes allow excessive ultraviolet radiation from the sun to reach the Earth. To date, the amounts have been small and pose no direct threat to large animal life. However, this radiation also impedes photosynthesis. It could well affect the growth of plankton, the basis of all polar food chains.

Antarctica is rich in mineral deposits. While no major efforts to tap this resource are underway, such a project could devastate the region. Mining operations require massive logistical support, which would mean assuming control of the little habitable land the region has. Accidents like oil spills along with the normal pollution that accompanies commercial efforts are other concerns.

Like all wildlife, polar animals do not live an isolated existence. Many species migrate to warmer seas during certain seasons or certain stages of their lives. Some polar species are staple food items for animals that inhabit temperate and tropical regions. Others help to keep these regions in balance, preying on creatures and keeping their populations in check. The polar seas might seem isolated and far from the rest of the world, but their inhabitants are very much a part of the global ecology. How we deal with the Arctic and Antarctic will likely affect much of the world.

Right: *Polar bears are solitary animals by nature; adults normally gather only to mate. The exception is a mother and her cubs, who will spend almost every minute of two years in each other's company.* Opposite page: *What appears to be a fierce dispute is actually a pair of cubs at play.*

LIFE IN POLAR REGIONS

LIFE IN POLAR REGIONS

continued from page 285

Various seals and walruses inhabit polar seas. In the Arctic and sub-Arctic, harp seals, ringed seals, bearded seals, ribbon seals, gray seals, and walruses are the most prominent species. Harp seals are migratory animals, ranging far north in the summer to hunt herring, cod, capelin, squid, and crustaceans. As winter and harsh weather approach, these seals travel south.

Small and handsome ringed seals are one of the Arctic's most common seals. Their wide range covers all corners of the northern polar regions. They often burrow tunnels in the ice to protect themselves from the cold.

Bearded seals are also wide ranging, but unlike many seals, they are often solitary. They take up residence on moving sea ice and spend much time on or near ice floes in shallow waters. They make extensive dives to scour the sea bottom for mollusks and crustaceans.

Known for their tusks, walruses are among the largest pinnipeds in Arctic waters. Large males reach sizes of 11 feet and 3,500 pounds. They are highly gregarious animals and often form herds of a thousand or more. Walruses hunt in shallow water and almost never stray far from coastal areas.

The Antarctic is home to four seal species. Crabeaters are the most abundant. Despite their name, crabeaters feed on plankton. Their cheek teeth have very elaborate cusps made for straining krill and other planktonic organisms from the water. Weddell seals inhabit the permanent, or fast, ice along the Antarctic continent. Superb divers, they go as deep as 1,200 feet in search of fish and squid.

Opposite page: *A seal pokes it head up through a hole in the ice to breath. Luckily, no polar bear is lying in wait.* Below: *A group of crabeater seals rest in the snow of Antarctica's Paradise Harbor.*

LIFE IN POLAR REGIONS

This page, top: *Rolls of blubber under a fur coat help keep seals warm in the chill of the polar air.* Above: *After a quick breather at a hole in the ice, a crabeater seal returns to its hunt for food.* Opposite page: *An arctic tern guards its pair of mottled eggs. The eggs' coloration makes it difficult for predators to see them on Alaska's rocky beaches.*

Leopard seals are solitary animals that normally gather only to breed. Wide ranging, they sometimes visit the coasts of New Zealand, Australia, and South America. They relentlessly prey upon fish, birds, and other seals, and their varied diet affords them such a broad distribution. Ross seals are the least common Antarctic seals. They are solitary animals and have not been well studied.

During the summer months, the polar and subpolar coasts host tens of thousands of seabirds. Guillemots, gannets, razorbills, gulls, shags, cormorants, and many other species use the narrow ledges and outcroppings of sheer coastal cliffs as nest sites. These species depend upon fish, crustaceans, and mollusks for their food. Some swoop down and pluck creatures from surface waters, while others dive hundreds of feet to find food.

Antarctica has only about 40 major bird species, including albatross, cormorants, ducks, terns, and penguins. The variety is limited, but many species have huge populations. Most use the islands and any exposed lands for breeding.

These birds normally colonize the coastal areas of sub-Antarctic waters. When summer eases the polar climate, many seabirds migrate south toward the abundant food supplies. As winter approaches, they seek the north's warmer weather. One species, the Arctic tern, annually migrates from the Antarctic to the Arctic, covering more than 18,000 miles.

LIFE IN POLAR REGIONS

Penguins are among the most unusual Antarctic birds. They are perfectly suited for their aquatic lives. Their wings have evolved into fins, they have webbed feet, and their thick feathers keep them warm in chilly seas. Like marine mammals, penguins have a layer of fat that adds warmth and provides a reserve food supply.

This page, top: *Two gentoo penguins guard an egg on the Antarctic shore. Their nesting behavior typically includes a good deal of ritual squawking and screeching.* Above: *Penguins often rest on the snowy landscape by either standing or laying on their bellies.* Opposite page: *A numberless mass of penguins crowd this snowy bank, each waiting patiently for its turn to enter the water and feed.*

Penguins feed at sea on fish and mollusks, but little else is known of their seaward excursions. Most study focuses on the time they spend on land where they hatch and raise their young. All 17 penguin species live south of the equator, and many inhabit the Antarctic. The Antarctic and sub-Antarctic species include gentoo, chinstrap, macaroni, rockhopper, king, emperor, and Adélie penguins.

Emperor penguins have a deliberate, seemingly regal way of walking. They are fascinating birds. The largest of all penguins, full-grown emperors stand four feet tall and weigh more than 90 pounds. These amazing creatures breed in the open at the height of the Antarctic winter, facing perhaps the world's worst weather conditions with virtually no shelter. They go about the business of raising their young amid temperatures of 80 below and howling 100-mile-an-hour winds.

A female emperor penguin lays a single egg. She leaves the task of incubating the egg to her mate, departing for sea where she will feed for the next two months. The male gently carries the egg on top of his feet, covering it with a fold of abdominal skin. He will not leave the egg to feed for the entire two-month incubation period; a male can lose half his body weight before the egg hatches.

About the time the chick emerges, the fattened female reappears. The male gladly surrenders his charge and heads to sea to replenish his own reserves.

LIFE IN POLAR REGIONS

LIFE IN THE OPEN OCEAN

For many open ocean creatures, the environment offers no accessible bottom, no surfaces that life can cling to. It has only water—vast and everywhere the same. It lacks any obvious sanctuaries, any refuge from its uniformity. There is no place to rest and no place to hide. Still, hundreds of species thrive here, endlessly cruising the ocean's surface waters. Home is wherever they happen to be.

In many areas of the open sea, marine life may be abundant one day and absent the next. The creatures that live here are transient and migratory, always on the move. Scientists cannot predict where any animals will be and when they will be there. Over time, though, they have begun to develop an accurate picture of life in the open ocean.

In some locations, the presence or absence of life is somewhat steady. For example, there is relatively little life in the Sargasso Sea, an area in the western Atlantic bounded by ocean currents. Also, surface life often correlates directly to bottom terrain even though the seafloor is hundreds

Opposite page: *Omnivorous saucereye porgies may become prey to larger fish as part of the ocean's food web.* Above, left: *A hammerhead shark, a large fish at the top of the food web, has few natural enemies.* Above: *The purple float of a Portuguese man-of-war is carried by the currents and pushed far out to sea by the wind.*

LIFE IN THE OPEN OCEAN

of feet deep. Different terrains attract different bottom dwellers. They in turn attract different predators that prowl the water directly above them, and so on to the surface. Dwarf sperm whales, for example, live exclusively over continental shelves because they can fill their diet only in these areas. They dive in the deep waters just past the shelf edge to hunt for squid and search the shallower shelf floor for fish and crabs.

As in most marine settings, the open ocean food chains rest on plankton, microscopic plants and animals that drift through the seas. Large concentrations range from only a few square yards to a few hundred square miles. Some zooplankton live in the open sea permanently. Other forms are the larvae of fish, lobsters, and other animals that reside in different habitats as adults.

Many boaters and divers rarely notice the small planktonic creatures except when dense populations alter water color or reduce underwater visibility. Marine creatures, though, pay close attention to the plankton. The plankton supply directly influences the creatures that feed on plankton and the creatures that feed on the plankton eaters. Without plankton, life in the open sea cannot survive.

Several factors influence the distribution of plankton. Currents are one of the most important considerations, but other things play a part as well. Temperature, wind, tide, and storms all affect the plankton.

Flourishing plankton attracts many species, including herring, anchovies, sar-

Opposite page: *Millions upon millions of baitfish school in the South China Sea. In this massive silvery cloud, their motion may confuse smaller predators.* This page, top: *Atlantic bottlenose dolphins cruise shallow waters near shore as well as open ocean waters. In coastal waters, they usually feed on bottom-dwelling fish, but they snap up active swimming fish when out to sea.* Above: *Humpback whales also spend much time near shore, but they are comfortable in the remotest parts of the Atlantic and Pacific, too.*

LIFE IN THE OPEN OCEAN

This page, top: *Many squid swim in schools and have no trouble grouping to spawn, but the solitary animals in the open ocean have more trouble finding mates.* Above: *Some very large animals, like the manta ray, are filter feeders, depending on small plankton for food.* Opposite page: *Under the surface, the long tentacles of a Portuguese man-of-war trail until they catch a fish. The man-of-war can handle some surprisingly large prey.*

dines, shad, and other fish. These creatures attract larger predators—bonito, yellowtail, barracuda, albacore, and jacks. And the mid-sized predators draw still larger hunters, such as swordfish, sailfish, marlin, dolphins, sharks, and toothed whales. All these creatures constantly travel the great tracts of open ocean, tirelessly seeking food and fleeing their predators. They are truly the nomads of the open sea.

ANIMALS OF THE OPEN OCEAN

Many open ocean species are actually poor swimmers. Most of these are invertebrates, such as jellyfish, Portuguese man-of-wars, sea wasps, and other similar creatures. Unable to effectively control their own movement, they go where the prevailing winds and currents take them. Because they lack the mobility to pursue food, they rely on the potent stinging cells found mainly in their tentacles. When they happen upon an item of prey, the cells automatically release deadly toxins that incapacitate the animal.

The potency of these toxins varies widely from one species to the next. The moon jelly, an attractive Caribbean jellyfish, is comparatively harmless to humans. Its stinging cells cannot penetrate human skin. Box jellies and sea wasps can inflict fatal stings. Relatives of the jellyfish, box jellies and sea wasps have four main tentacles that attach to their bell. These tentacles often branch, forming a network that looks like the roots of a tree. The most potent sea wasp inhabits Australian waters, while many less potent species wander across the other tropical seas.

LIFE IN THE OPEN OCEAN

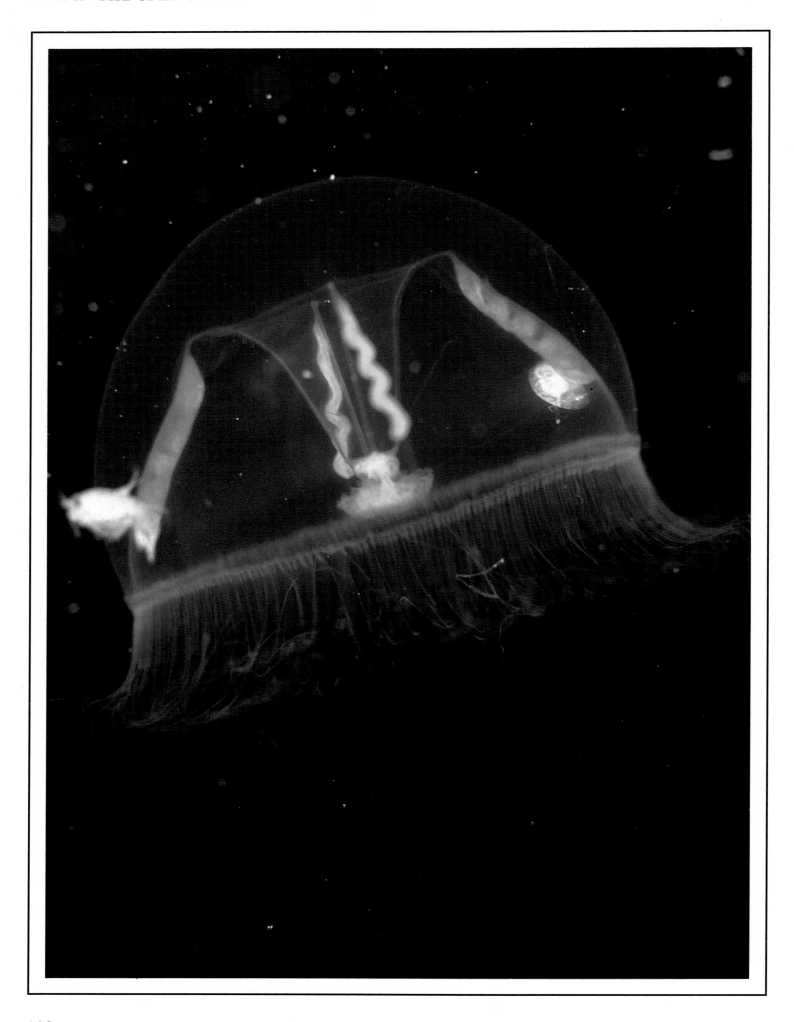

The comb jelly is another open ocean invertebrate similar to the jellyfish. These drifters have distinctive shapes and eight rows of hairlike cilia that help them maneuver. The rows of cilia look similar to the teeth of a comb. Comb jellies come in various shapes, but none have the typical bell of a true jellyfish.

Another well-known open ocean animal is the Portuguese man-of-war, which populates tropical and semitropical seas. From the surface, its blue to purple bell is often evident as it rides the waves. The bell is generally less than a foot across, but the powerful, nearly transparent tentacles often trail 30 feet or more behind the bell.

The man-of-war fish is a small open sea creature that sometimes hides among the man-of-war's tentacles. The fish gain protection by being able to hide, but they may not be immune to the tentacles' deadly toxin. Vulnerable to predators should they leave the Portuguese man-of-war and perhaps just a careless moment away from becoming its prey, these small fish truly lead a dangerous life.

A number of animals can eat jellyfish despite their stinging cells and toxins. They are a favorite food of the strange-looking ocean sunfish. Ocean sunfish have distinctly flattened bodies. They look somewhat like enormous frisbees with fins, and they can grow to be 3,000 pounds. Ocean sunfish are usually solitary, but they sometimes travel in small schools. Sea turtles also readily prey upon jellyfish, both in their shallow reef communities and as they migrate through the open sea.

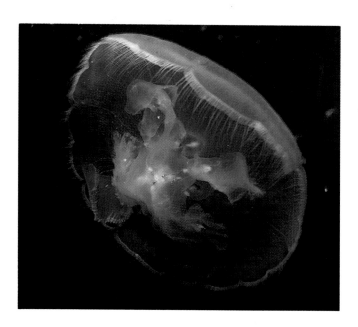

Opposite page: *Jellyfish come in all shapes and sizes, but they all have a gelatinous bell and a fringe of stinging tentacles. The length of the tentacles and the potency of their sting varies from species to species.* This page, top: *Although some jellyfish are almost transparent, species such as this one from the South Pacific are colorful and appear more solid.* Above: *The moon jellyfish, common in the Atlantic Ocean, has a very short fringe of tentacles.*

LIFE IN THE OPEN OCEAN

This page, top: *A scuba diver is dwarfed next to an ocean sunfish. These large fish feed mostly on jellyfish, and occasionally on fish, algae, and invertebrates.* Above: *Although turtles are often seen near reefs, they also venture out into the open ocean.* Opposite page: *This daisy chain of salps looks like a string of large ocean jewels. These gelatinous animals are related to the tunicates that inhabit reefs and rocky shores.*

The Importance of Sunlight

Sunlight is as important to marine life as it is to life on land. Sunlight fuels the process of photosynthesis in marine plants, which serve as the foundation for most marine food chains. The vast majority of ocean creatures live in water shallow enough for sunlight to penetrate, even though this is only about two percent of the ocean's water.

Sunlight regulates the daily cycle of many animals. Many forms of zooplankton and other open sea invertebrates migrate toward the surface at night to feed and then descend with the rising sun. Other animals have active and inactive periods during the day that begin and end with cues from the sun.

The sun is also an important factor in the migratory cycles of some ocean creatures. When sunlight encourages plankton blooms, the great whales head toward their feeding grounds. When the blooms end, they swim to their breeding areas. The sun's rays may also help many fish and other marine animals stay on course as they migrate.

Salps are another unusual open ocean specimen. These semitranslucent, gelatinous animals are something of a link between vertebrates and invertebrates. Some are solitary and grow to the size of a human fist. Others form linked chains, usually a few feet long but sometimes measuring 60 feet. Poor swimmers, salps drift in the currents of the open sea.

LIFE IN THE OPEN OCEAN

LIFE IN THE OPEN OCEAN

LIFE IN THE OPEN OCEAN

Perhaps the most varied open ocean dwellers are the fish. Some wander aimlessly from this place to that, seeking their food and then moving on. Others have clear migratory patterns based on seasonal cycles. Traveling alone or in huge groups, they continuously patrol the ocean's expanse.

Worldwide there are about 140 species of jacks. They are swift, usually silver-colored fish that inhabit the open ocean and some tropical and temperate reef communities. Many jacks use their lightning speed to strike small schooling fish.

Yellowtail are large fish that form schools in the surface waters of the open ocean. They often grow to well over 20 pounds. They feed on squid, anchovies, crabs, sardines, mackerel, and other open sea creatures, and several sharks prey on the yellowtail.

Found in temperate and tropical seas of the north and south, bonito are fast predators that school in the thousands. These silver to blue, medium-sized fish excite easily and often swirl wildly around anything that happens to catch their attention.

Flying fish are among the most fascinating open sea creatures. Inhabiting tropical and temperate seas around the world, the 50 species of flying fish can glide above surface water to evade dolphins, fish, and other potential predators. To become airborne, they beat their tails rapidly to gain speed as they approach the surface. Once they breach, flying fish spread their modified pectoral fins. The large fins act like the

Opposite page: *A large school of jacks form in a swirling mass.* This page, top: *Flying fish usually dwell in the surface waters of the ocean so that the safety of taking to the air is never far away.* Above: *Bonito have the classic torpedolike body shape that reduces water resistance as they swim. They are built for speed, and they use it both in hunting and in escaping predators.*

This page, top: *Sailfish fold their high dorsal fin down into a groove on their backs when swimming at great speeds. They are thought to use their large dorsal fin to round up schools of baitfish prior to feeding.* Above: *Striped marlin are excellent game fish, known for putting up long, hard fights.* Opposite page: *Tuna school in the open sea. Tunas and the related mackerels are the most streamlined of all fishes.*

wings of an airplane, creating lift that supports them as they skim over the waves. They can reach 35 miles per hour and glide 150 feet or more.

Marlin and sailfish are among the many predators that hunt flying fish. The long, thin extension of their upper jaws has earned them the name billfish, and it has also given them a distinct advantage in the open sea. The bill helps to increase the fish's swimming speed by reducing water resistance. It works much the same as the tapered nose of an airplane or missile, and it works quite well. Billfish are among the fastest open ocean species.

Billfish are a rather advanced group, well adapted to pelagic, or open ocean, life. They are active hunters in tropical and warm temperate seas, feeding on fish, squid, and crustaceans. Weighing more than 2,900 pounds, the black marlin is the world's largest bony fish. By comparison, the blue marlin can weigh 1,200 or 1,300 pounds. Striped marlin, white marlin, and sailfish are a good deal smaller.

Tuna, another fast open ocean species, are highly migratory. They show several adaptations that support their nomadic lifestyle. A tuna's circulatory system is unlike those of most other fish. The circulatory system can retain a limited amount of body heat as needed. Unlike most other bony fish, tuna maintain a temperature a little higher than that of the surrounding water. The higher temperature gives them a slight edge, enhancing their strength, speed, and endurance. Sometimes, it's just enough to escape one of their many predators.

LIFE IN THE OPEN OCEAN

LIFE IN THE OPEN OCEAN

This page, top: *Japanese barracuda, schooling in the ocean off Hawaii, don't grow as large as the barracuda that inhabit the inshore waters of the Atlantic.* Above: *A Pacific manta ray, the largest of the mantas, can have a wingspan of over 20 feet and may weigh over 3,000 pounds.* Opposite page: *Blue sharks are truly pelagic creatures. They almost never enter waters less than 100 feet deep.*

To maintain their high temperatures, tuna eat huge amounts of food. In cool waters, some species consume 10 percent of their own weight every day. Although they normally inhabit the ocean's surface waters, tuna do sometimes dive to considerable depths as they pursue food and try to avoid predators.

Seamounts

Seamounts are towering undersea mountains that rise dramatically from the ocean's depths. Most have their origins in volcanoes or other violent geological activity. These undersea pinnacles stand firmly in the open sea, bathed by currents and nutrient-rich upwellings.

The sheer walls, crevices, ledges, caves, and plateaus of seamounts create ideal living quarters for many animals. They provide a solid surface that sponges, barnacles, corals, scallops, worms, and anemones can use to secure themselves. Other invertebrates from sea stars and octopi to lobsters and crabs find seamounts to their liking. Their presence attracts many predators. Manta rays and turtles often congregate at seamounts, as do many schooling fish such as jacks, bonito, tuna, yellowtail, and mackerel that normally roam the open sea. Of course, their presence brings dolphins, billfish, barracuda, sharks, and other large open sea predators.

Although seamounts are small when compared to the vast open sea, these undersea mountains are usually a hub of marine activity.

LIFE IN THE OPEN OCEAN

LIFE IN THE OPEN OCEAN

LIFE IN THE OPEN OCEAN

Many sharks play important roles in the open sea. Reaching a maximum length of almost 40 feet, the basking shark is the second largest shark species. Only the whale sharks of tropical seas are larger. Like whale sharks, basking sharks are filter feeders; their diet consists mostly of plankton. Other open water sharks, such as makos, blues, and threshers, feed on open sea fish and squid.

In tropical seas, oceanic whitetips fill a niche very similar to that of temperate water blue sharks. These whitetips are impressive-looking sharks that have a reputation for being bold. Pilot fish sometimes accompany the whitetips, feeding on scraps after the shark makes a kill.

Scalloped hammerhead sharks gather near seamounts during the day, but they often swim in open water at night when they hunt squid and fish. Scientists are still trying to learn how these sharks can find their way back to the seamounts after making long nighttime trips.

Remoras are fish that attach themselves to sharks and other large sea creatures to obtain both protection and food. Remoras clean sharks of parasitic crustaceans. They occasionally end up as a meal for the shark, but they do gain protection from other predators. Remoras often attach to manta rays, porpoises, turtles, and other nomadic creatures as well.

Several species of dolphins and whales commonly swim in surface waters many miles from the nearest coast. Common dolphins often dive 800 feet or more as they pursue fish and squid. Spinner dolphins often asso-

Opposite page: *Blue sharks are known to dive to great depths, but they spend most of their time stalking the waters near the surface.* This page, top: *An oceanic whitetip shark is accompanied by a group of pilot fish. The pilots scavenge leftovers from the shark's meals.* Above: *Remoras attach to other creatures using the raised, slightly ridged sucker disk atop their heads.*

LIFE IN THE OPEN OCEAN

ciate with migrating yellowfin tuna. Risso's dolphins, bottlenose dolphins, striped dolphins, and white-sided dolphins are other open ocean species. Some of these also frequent shallower coastal waters.

Pods of sperm whales and pilot whales often travel the open sea. The sperm whales frequently make deep dives for giant squid and other food. Other whale species traverse the open sea at certain times of year as they migrate between breeding and feeding areas.

As with other major marine habitats, the open sea is home to a great many interconnected species. Wandering alone or in huge schools, these creatures support and rely on the other members of their watery environment.

Above: *Humpback whales make long migratory excursions across oceans from their temperate feeding grounds to tropical and subtropical calving and mating grounds.* Opposite page: *Acrobatic spinner dolphins often make spectacular leaps from the water. They often travel with tuna, perhaps because they both feed on the same small fish and squid. Spinner dolphins travel in large pods of thirty to several hundred animals.*

Crossing the Great Expanses of Ocean

Fish ranging from mackerels to tuna to eels undertake annual migrations that cover two, three, four, even five thousand miles. Often these trips take fish from feeding grounds to breeding grounds and back again in an ongoing cycle. Several theories account for the fish's ability to navigate, but as yet there is no definite explanation of how they cross the vast, almost featureless ocean without getting lost.

Migratory patterns often correlate strongly with ocean currents. In some cases, the fish swim with the currents and in other instances they oppose the currents. They may be using these strong water channels as roadways that lead to their destinations.

The angle of the sun's rays may also offer the travelers some guidance. At different times of year, the sunlight's angle of penetration into the water varies. Some migratory species might instinctively measure the angle of the sun's rays and use this information to follow a specific course.

Some studies have shown that certain fish have an internal compass of sorts. Some migratory fish have small particles of the mineral magnetite in their bodies. The earth's magnetic fields are fairly constant, and the magnetite may allow them to key on the earth's poles as guideposts.

LIFE IN THE OPEN OCEAN

Some residents of the Sargasso Sea: This page, top: *A small sargassum shrimp will spend its entire life moving in and around a clump of sargassum weed. This brown algae has gas-filled bladders that keep it at the ocean's surface.* Above: *The ornate coloration and frilly appendages of the small sargassum fish keep it well camouflaged amid the tangle of weeds.* Opposite page: *Not all the animals attracted to the sargassum are permanent residents. A freshwater European eel will venture out to the Sargasso Sea to spawn. Its young will eventually return to fresh water to live out most of their lives.*

The Sargasso Sea

Unique conditions have created a very unusual setting in the Atlantic waters east of Bermuda. The 2,000-square-mile Sargasso Sea is bounded by several major ocean currents including the Gulf Stream. These powerful currents isolate the area from surrounding waters, and the region's climate produces meager rainfall and heavy evaporation. The combination of isolation and climate creates warm water with a very high salt content.

The Sargasso Sea is relatively devoid of life. The waters contain very few basic nutrients. Plankton production is low, so many higher animals cannot survive there. But even in these bleak surroundings, some species do manage to flourish. Sargassum weed, which lends the area its name, is the most abundant life-form. About 10 million tons of this floating weed grow there every year.

The plants grow together in tangled rafts that provide shelter and food for some very hardy animals. The sargassum fish, for example, is an expertly camouflaged fish that uses a lure on its dorsal spine to attract unwary prey. Eels such as the North American common eel and European freshwater eel use the region as a spawning ground. The hatchling eels may spend a year or more wandering the ocean before they find their freshwater homes.

LIFE IN THE OPEN OCEAN

315

Glossary

Attenuated Having an elongated, snakelike body shape. Examples of attenuated fish are eels, needlefish, and pipefish.

Baleen A series of horny plates in the mouths of filter-feeding whales that strain plankton and small fish from the water.

Budding A form of asexual reproduction in which part of the parent organism pinches off and forms a new organism. Some jellyfish, anemones, and corals reproduce by budding.

Buoyancy The ability of a body to float when submerged in water because the body's density is less than that of the surrounding water.

Carnivore An animal that eats other animals. Examples of marine carnivores are sharks, toothed whales, barracuda, and some sea stars, anemones, and worms.

Cetacean A taxonomic order of marine mammals that includes whales, dolphins, and porpoises.

Chromatophore A pigment cell usually found in the skin that can be contracted or expanded to change color. Octopi and fish use them for camouflage and communication.

Cilia Short, hairlike appendages that can be used for movement or capturing prey. Some worms and sponges have cilia.

Cirri Slender, flexible appendages used by suspension feeders or filter feeders such as barnacles and feather stars to capture food.

Cnidocytes The stinging cells found in corals, anemones, jellyfish, and hydroids.

Colony Individual animals that are joined together and function as a single unit. Examples of colonial marine animals are Portuguese man-of-wars and some hydroids.

Commensalism A relationship between two organisms in which one benefits and the other neither benefits nor suffers.

Current A steady flow of water within the ocean along a relatively constant path, much like a river within the ocean.

Deposit feeder An animal that feeds by foraging for tiny organic particles that have settled on the seafloor.

Diurnal Having a daily cycle of active and inactive periods.

Dorsoventrally compressed Having a body that is much wider than it is tall. Examples of dorsoventrally compressed fish are rays, turbot, and flounder.

Echolocation A sensory skill in which an animal emits sound waves and then collects and studies the reflected echoes. Toothed whales use echolocation in hunting.

Filter feeder An animal that feeds by straining organic particles or small organisms from the water.

Fusiform Having a tapered, torpedolike body shape. Examples of fusiform fish are great white sharks and barracuda.

Habitat A specific place or set of environmental conditions in which an organism normally lives. Major marine habitats include coral reefs, kelp forests, and polar seas.

Herbivore An animal that eats plants. Examples of marine herbivores are parrotfish, manatees, and some snails and sea urchins.

Hermaphroditic Having both male and female reproductive organs. Examples of hermaphroditic marine animals are sponges and some snails.

GLOSSARY

Holdfast   The part of a kelp plant that anchors the plant to the seafloor.

Invertebrate   Any animal that does not have a backbone. Examples of marine invertebrates are corals, octopi, worms, and sea stars.

Laterally compressed   Having a body that is much taller than it is wide. Examples of laterally compressed fish are angelfish and triggerfish.

Mutualism   A relationship between two organisms in which both benefit.

Nematocyst   A structure within the stinging cells of corals, anemones, jellyfish and other cnidarians that is fired into other animals. Nematocysts function primarily as a means of hunting and defense.

Operculum   A bony covering over the gills of many fish that pumps water over the gills.

Parasitism   A relationship between two organisms in which one benefits and the other is harmed.

Pedicellariae   Small, pincerlike organs in some echinoderms that are used in defense and hunting.

Pelagic   Living in the open ocean. Examples of pelagic animals include blue sharks, swordfish, and some squid and crabs.

Photosynthesis   A process in which plants use sunlight and inorganic compounds to produce carbohydrates for energy.

Pinniped   A taxonomic order of marine mammals that includes seals and walruses.

Plankton   Small, often microscopic, plants and animals that float freely in the ocean. Plankton forms the basis of almost all oceanic food chains.

Radula   A horny, tonguelike appendage that many snails and some other mollusks use to gather their food.

Setae   Slender, rigid, sometimes sharp appendages of some segmented worms used for movement and defense.

Spawning   The reproductive method of most fish and some marine invertebrates in which the males and females release their gametes, or sex cells, into the water and fertilization occurs outside the body.

Surf   Ocean waves that break on the shore.

Surge   A large, heavy wave or series of large heavy waves beneath the water's surface.

Suspension feeder   An animal that feeds by capturing tiny organic matter as it drifts through the water.

Swell   Surface waves or a series of surface waves away from shore.

Symbiosis   Any close relationship between two organisms.

Symmetrical   Consisting of two or more parts that are nearly identical in structure and function.

Terrestrial   Living on land rather than in water.

Trophic   Having to do with food chains or the relationships between organisms and their food sources.

Upwelling   The movement of water from any depth toward the surface.

Vertebrate   Any animal that has a backbone. Examples of marine vertebrates include fish, turtles, seals, and whales.

Index

A
Abalone, 97
Algae, 238
Anemonefish, 27, 80, *80-81*
Anemone, 27, 58, *60*, *63*, *78-79*, 79-80, *255*, *264*, *265*, *266*, *270*
Angel shark, *160*, 183, *183*, 269
Angelfish, *31*, *124*, *154*, 244
Annelid worms. *See* Segmented worms.
Antarctic, 18, 273-293
Arctic, 199, 273-289
Arctic tern, 290, *291*
Arthropods, 104-111
Australian flatback green turtle, 133

B
Baitfish, *155*, *296*
Barnacles, 110, *110*, 263, 273
Barracuda, *10*, *32*, 48, *48-49*, 247, *264*, *308*
Basket sponge, 66
Basket star, 113, 116, *117*
Basking shark, 311
Batfish, *34*
Bearded seal, 197, 289
Beluga whale, *214-215*, 283
Bering Sea, 206
Billfish, 306
Black coral, *72*, 73
Black grouper, *20*
Blacktip shark, *159*
Blonde sea lion, 194, *194*
Bloodworm, 90
Bloody Bay Wall, 231
Blue chromis, *225*
Blue crab, 107
Blue shark, *166*, 168, *174*, 179, *179*, *309*, *310*
Blue tang, *237*
Blue whale, *211*, 213, 283
Blue-ringed octopus, 61, *61*, *102*
Bonito, 305, *305*
Bottlenose dolphin, *38*, *204*, *297*
Bowhead whale, 283
Box jellyfish, 298
Bristle worm, *57*, *58*, 90, 244
Brittle star, 12, *12*, 113, 116, *116*, *263*, *279*
Bryozoans, *92-93*, 93
Butterfly fish, *8*, 16, *34*, *141-142*, *222*, *235*
By-the-wind sailor, 85, *85*

C
California gray whale, 206, *207*
California moray eel, *238*
California sea lion, *9*, *190*, *192*, *193*, 193-194, *253*
California spiny lobster, *29*, 106
Camouflage, 102-103, 107, 142, 227, 271
Cardinal fish, 244
Caribbean, 64, 71
Caribbean creole fish, *23*
Caribbean king crab, *245*
Caribbean spiny lobster, 106
Carnivores, 21
Cephalopods, 102-103
Cetaceans, 37-38, 205-221
Chondrichthyes, 138
Christmas tree worm, *88*, 89
Chromataphores, 103, 137
Clams, 25, 100, *100*, 269
Cleaner shrimp, 22, *26*, 108, *108*, *238*
Cleaner wrasse, 22, *22*, 24
Cleaning relationships, 22
Clown triggerfish, 140
Cnidaria, 68-86
Coccina clams, *100*
Comet fish, 15
Commensalism, 27
Conchs, *95*, 97
Cone snail, 58, 97
Copepod, *28*, 111
Copper rockfish, *269*
Coral, 24, 58, 71-77
Coral reefs, 75, 223-247
Countershading, 183
Cownosed ray, *187*
Cowries, 97, *97*
Crabeater seal, 289, *289*, 290
Crabs, 24, 107, *107*, 244, *245*, 263, 269, 273
Creole fish, 22, *23*
Crinoids, 122, *122*, *123*
Crown-of-thorns sea star, *237*

Crustaceans, 104-111
Cup coral, *229*
Cuttlefish, 102, *102*

D
Damselfish, 240, *241*, 244
Dogfish, *169*
Dolphins, 16-17, 37-38, *38*, 189, *204*, 205, 216, 218, 221, 311, 312
Dungeness crab, *274*
Dwarf sperm whale, 297

E
Eagle ray, 184, *185*, *240*
Echinoderms, 113-123
Echolocation, 16, 50, 215, 221
Elasmobranchs, 158
Electric ray, 187, *187*, 269
Elephant seal, 196-197, *196*, *197*, 280
Elkhorn coral, *71*
Emperor penguin, 292
European eel, *315*
Ewa blenny, *246*

F
Fairy basslet, *225*
False cleaner, 24
Feather duster worm, 90
Feather star, 115, 122, *122-123*
Filefish, *154*
Fire worms. *See* Bristle worms.
Fish, bony, 137-155
   coloration in, 142
   senses in, 140-141
   sleep in, 141
Fish, cartilaginous, 138, 157-187
Flame scallop, *101*
Flamingo tongue cowrie, 97
Flatworms, 86-89, *87-89*
Flying fish, *305*, 305-306
Food chains, 18-21, 75, 238, 283
French angelfish, *239*
Fungus coral, *71*

G
Galápagos Islands, 129, 193
Galápagos shark, 180, *180*
Garden eel, *227*

INDEX

Garibaldi, *251*, 254, *264*
Gastropods, 97
Gentoo penguin, *292*
Ghost pipefish, *139*
Giant clams, 24, *25*, 100
Giant kelp, *249*, 250, 252, *252-254*, 254, 257
Giant squid, 102
Glass fish, *12*
Glass jellyfish, *273*
Goatfish, 12, *148*
Goby, 22, 27, *27*
Goose barnacle, 110, *110*, 262
Gray reef shark, 180, *180*
Gray seal, 289
Great Barrier Reef, 235, 237
Great white shark, *13*, *42*, 44, *45*, 47, *170*, 172, 174-177, *175-176*, 267
Green tube sponge, *236*
Green turtle, 133-134, *133*
Greenland, 197
Grey whale, *283*
Grouper, 22
Grunt, *237*
Guitarfish, 269

H
Hammerhead shark, *157-158*, *178*, 179, *295*, 311
Harbor seal, *190*, 197
Hard coral, *228*
Harp seal, 197, 289
Hawaiian lobster, *104*
Hawksbill turtle, 133
Heart urchin, 118
Herbivores, 19
Hermit crab, 27, 107, *107*, 244
Hooded seal, 197
Hooker's sea lion, 194
Horn shark, *47*, *47*, *182*, 183, 267
Horse conch, 95
Humpback whale, *205*, *208-210*, 209, 210, 213, 283, *297*, *312*
Hydroids, 58, 85, *85*, 268

I
Icefish, *279*
Imperial shag, *278*
Invertebrates, 62-123
Isopods, 111, *111*

J
Jacks, 16, *33*, *147*, 247, *304*, 305
Japanese barracuda, *308*
Jellyfish, 16, 58, 82, *82-83*, 298, 301, *300-301*
Jellyfish medusa, *68*
Jellyfish, comb, *82*
Jellyfish, moon, 82, *82*

K
Kelp, 250, 252, *252*, *253*, 254, *254*, 257-258, 260
Kelp crab, *257*
Kelp snail, *18*, *256*
Kelpfish, *249*, 250, 257
Kemp's ridley turtle, 133
Killer whale, *44*, 50, *50*, *51*, *189*, 205, 215-216, *216-217*, 283
King angelfish, *136*
King crab, 107
King penguin, *272*
Krill, 111, 280, *280*

L
Leatherback turtle, 133-134
Leopard seal, 53, *194*, 290
Leopard shark, *167*
Lionfish, 54, *56*, *57*, *143*
Little Cayman Island, 231
Lizardfish, *137*, 138
Lobsters, 106, *106*, 263, 273
Loggerhead turtle, *133*, 134

M
Mako shark, *177*
Mammals, 17, 189-221
Manatee, 189, *202*, *203*, 203
Manta ray, 184, *184*, 187, *298*, *308*
Mantis shrimp, 108
Marbled ray, *156*
Marine iguana, *129-130*, 129-131
Marlin, *306*, *306*
Marshall Island goby, 145
Masking crab, *24*
Micronesia, 243
Migration, 206, 312
Mimic blenny, 24
Minke whale, 283
Mollusks, 94-103
Moon snail, *96*

Moorish idol, *141*
Moray eel, *36*, 40, 44, 53, *53*
Mutualism, 22, 27, 75
Mysticeti, 205

N
Narwhal, 216, *282*, 285
Nassau grouper, *149*
Needlefish, *37*
Newfoundland, 197
Northern fur seal, *17*, *196*
Nudibranchs, 16, *16*, 86, 95, *95*, 98-99, *98-99*, 250, 263, 267
Nurse shark, *164*, 168

O
Ocean sunfish, 301, *302*
Oceanic whitetip shark, *165*, *173*, 311, *311*
Octopi, 14, 16, *21*, 41, *41*, 61, *61*, 63, 94, 102-103, *103*, 263
Odontoceti, 205
Olive ridley turtle, 133
Orange flame coral, 77
Orcas. *See* Killer whale.
Osteichthyes, 138
Oysters, 100
Ozone, 286

P
Pacific barracuda, *10*, *264*
Pacific octopus, *14*, 102
Pacific soft coral, *43*
Parasitism, 27-29, *28*, 111
Parrotfish, *11*, 240, 244, *244*
Peacock flounder, *15*
Penguins, 273, 276, 280, 281, 290, 292, *292-293*
Peru Current, 129
Philipine gorgonian coral, *73*
Phytoplankton, 18, 21
Pilot fish, 311, *311*
Pilot whale, 312
Pinnipeds, 190-199
Plankton, *19*, 237-238, 297
Polar bears, 273, *275*, *284-287*, 285-286
Polyps, 68, *76*, 77, *77*
Porpoises, 216, *221*
Portugese man-of-war, *11*, 54, *54*, *84*, 85, 298, 301, *299*

INDEX

Puffer fish, *144*
Purple sponge, *64*

R

Rays, *9*, 184-187
Red crab, *271*
Red rope sponge, *64*
Red sea fan, 69
Regeneration, 115, 121
Remoras, 311, *311*
Reptiles, marine, 129-135
Ribbon seal, 197, 289
Right whale, 39, 210-213, *212-213*
Ringed seal, 197, 289
Rock scallop, 100
Rockfish, *248*
Rockhopper penguin, *281*
Roundhead, 15

S

Sailfish, 306, *306*
Sally lightfoot crab, *105*, *107*
Salmon, *274*
Salps, 126, 302, *303*
San Miguel Island, 193
Sand dollar, 113, 118
Sand tiger shark, *167*
Sand tilefish, *33*
Sanddab, *138*, *269*
Sargasso Sea, 295, 314
Sargassum fish, 314, *314*
Sargassum shrimp, *314*
Saucereye porgies, *294*
Sawfish, *184*
Scallops, 100, *100-101*, 263
Schooling, 147, 148, 179
Scorpionfish, *43*, 54, 57, *57*
Sea bass, *233*
Sea biscuit, 118
Sea clown nudibranch, *8*
Sea cows. *See* Manatees.
Sea cucumbers, 113, *113*, *120-121*, 121
Sea dragons, 260, *261*
Sea hares, 98, *99*
Sea horses, *125*, 145, *145*
Sea lilies, 122
Sea lions, *17*, 189, 193-194, *194*, 273
Sea otter, 189, *189*, 200, *200-201*, 260
Sea pansies, 86
Sea pens, 86, *86*
Sea slugs, *250*
Sea snakes, *52*, 53, *125*, 131, *131*
Sea stars, *18*, 62, *112*, 113-115, *114-115*, 263, 273
Sea turtles, *128*, *132-135*, 133-134, *302*
Sea urchins, 58, *61*, 113, *113*, 118, *118-119*, *259*, 263
Sea wasps, 298
Sea whips, 75
Seals, 17, *17*, 189, 193-197, 273, 285, *288*, 289, *289*, 290, *290*
Seamounts, 308
Segmented worms, 89-91, *90*, *91*
Sei whale, 283
Sergeant major, *152-153*
Sharks, 11-12, 43-47, 157-183
 coloration in, 183
 reproduction in, 171
 senses in, 164, 168
Sheephead, 257
Shrimps, 22, *26*, *27*, 108, *108-109*, 244, 263, 269
Silverling, *146*
Silversided damselfish, *241*
Six-gilled shark, *174*
Skates, 184, *184*
Slipper snail, 97
Slipper lobsters, 106
Snails, 16, *96*, 97, 263
Snapper, *223*
Soft coral, 59
South American sea lion, 194
South Pole, 274
Southern stingray, *186*, 226
Spanish dancer, 98
Spanish hogfish, *232*
Spanish shawl, 98
Sperm whale, 196, 215, 312
Spider crab, *105*, *245*
Spinner dolphin, *218*, 221, 311-312, *313*
Spiral gill worm, *91*
Sponges, 16, 64-67, *64-67*, 263
Spotted dolphin, *218-220*, 221
Squid, 41, 102-103, *103*, 244, 273, *298*
Squirrelfish, *140*, 244, *247*
Starfish. *See* Sea Stars.
Stingray, 15, *30*, 54, *54*, *55*, 184, *186*, *187*, 269

Stonefish, 54, 57
Striped armina, *250*
Striped marlin, *306*
Sun stars, 115
Swell shark, *163*, *171*
Swim bladder, 30, 33
Symbiotic relationships, 24, 27

T

Tealia anemone, *255*
Territoriality, 148-151
Torpedo ray, *187*
Triggerfish, 22, *235*
Triton trumpet, *237*
Tropical soldierfish, *28*
Truk Lagoon, 243
Trumpetfish, 37, 138, *142*, *148*
Tuna, 306-308, *307*
Tunicates, 126, *126-127*
Turneffe Reef, 64
Turtles. *See* Sea turtles.

V

Vertebrates, 125-155

W

Walrus, *188*, 189, *198-199*, 199, 273, *273*, 289
Weddell seal, 194, *277*, 289
Whale shark, *46*, 47, *161*, 180, *181*
Whales, 17, 37, 38, *39*, 189, 205-217, 273
 baleen, 206-213
 toothed, 214-217
Whitetipped reef shark, *41*, *157*, *231*
Wire coral, 74
Wobbegong shark, *162*
Wolf eel, 264, 267, *267*
Wrasse blenny, 247
Wrasses, 16, 22, *22*, 24, *150-151*

Y

Yellow stingray, *35*
Yellow tube sponge, *65*
Yellowtail, 305

Z

Zooids, 93
Zooplankton, 18, 21, 105, 279, 297, 302
Zooxanthellae, 24, 75, 238

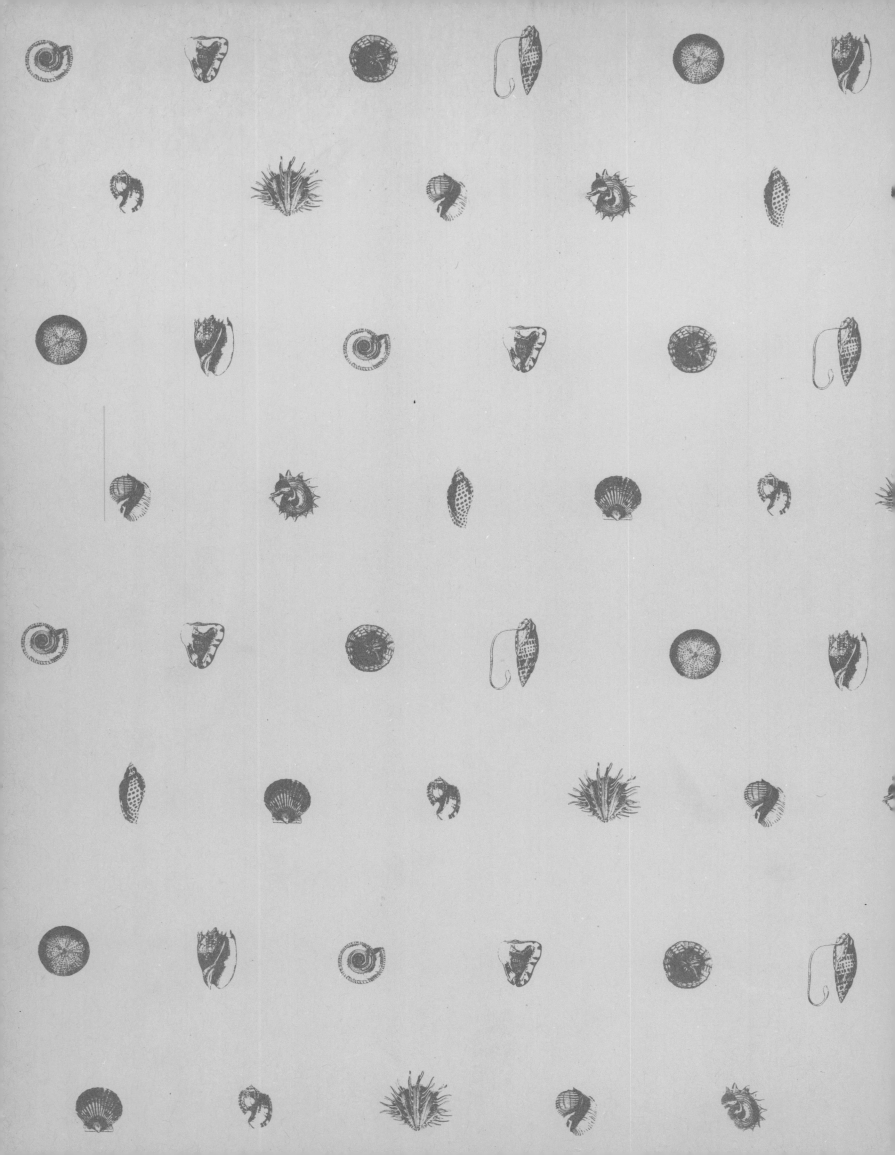